That's chemistry!
a resource for primary school teachers about materials and their properties

Compiled by Jan Rees
Teacher Fellow
The Royal Society of Chemistry
1998-1999

ROYAL SOCIETY OF CHEMISTRY

That's chemistry!

Compiled by Jan Rees

Edited by Colin Osborne and Maria Pack

Designed by Olley & Robertson

Published and distributed by The Royal Society of Chemistry

Printed by The Royal Society of Chemistry

Copyright © The Royal Society of Chemistry 2000

Registered charity No. 207890

Apart from any fair dealing for the purposes of research or private study, or critiscism or review, as permitted under the UK Copyright Designs and Patents Act, 1988, this publication may not be reproduced, stored, or transmitted, in any form or by any means, without the prior permission in writing of the publishers, or in the case of reprographic reproduction, only in accordance with the terms of the licences issued by the Copyright Licensing Agency in the UK, or in accordance with the terms of licences issued by the appropriate Reproduction Rights Organisation outside the UK. Enquiries concerning reproduction outside the terms stated here should be sent to The Royal Society of Chemistry at the London address printed on this page.

Notice to all UK Educational Institutions. The material in this book may be reproduced by photocopying for distribution and use by students within the purchasing institution providing no more than 50% of the work is reproduced in this way for any one purpose. Tutors wishing to reproduce material beyond this limit or to reproduce the work by other means such as electronic should first seek the permission of the Society.

For further information on other educational activities undertaken by The Royal Society of Chemistry write to:

The Education Department
The Royal Society of Chemistry
Burlington House
Piccadilly
London W1J 0BA

Information on other Royal Society of Chemistry activities can be found on its websites:
www.rsc.org
www.chemsoc.org
www.chemsoc.org/networks/LearnNet

ISBN 085404-939-8

British Library Cataloguing and Publication Data.

A catalogue for this book is available from the British Library.

FOREWORD

In recent years the amount of physical science taught in primary schools has increased dramatically and the importance of literacy and numeracy skills has been emphasised. The inquisitive nature of science is an exciting area for primary school children and the Royal Society of Chemistry is delighted to be able to support primary teachers by providing a resource that contains ideas, activities, demonstrations and investigations to enhance their teaching. Suggestions on how the children can access scientific ideas and methods via poetry accompanies each topic and will undoubtedly foster imaginative cross-curricular teaching. The resource concentrates on those areas of physical science that are related to materials and their properties and functions, and which may be less familiar to primary teachers from a non-science background.

This collection of material has been developed with the help and support of teachers throughout the United Kingdom. It is designed for both specialist and non-specialist primary teachers in the hope that they can engender interest and enthusiasm in the new generation of scientists.

Professor Steven Ley CChem FRSC FRS
President, The Royal Society of Chemistry

This resource was produced with support from the Royal Society of Chemistry 150th Anniversary Appeal Fund and the Association of the British Pharmaceutical Industry.

THE ROYAL SOCIETY OF CHEMISTRY

The Royal Society of Chemistry is the professional body for chemists and the learned society for chemistry. It is one of the most prominent and influential independent scientific organisations in Britain. Through its 46,000 members, including academics, teachers and industrialists, the Society promotes the interests of chemists and the benefits of chemical science.

The Society plays a leading role in the science of chemistry, communicating cutting edge research and its applications through highly respected journals and programme of international conferences, seminars and workshops. These meetings bring together chemists from all academic and industrial backgrounds and are overseen by the Society's six scientific Divisions. Affiliated to these Divisions are over 70 subject groups, each devoted to a specific area or application of chemistry.

The Society's educational activities provide information and training opportunities for both students and teachers enabling science and chemistry teachers to be continually involved with curriculum developments and abreast of the latest teaching methods and learning aids. Activities take place at both regional and national levels and the publications that arise from conferences and symposia form a valuable addition to the resources available to the teacher. The Society is extremely active in determining the future of chemical education, seeking to influence Government by submitting evidence to Parliament and anticipating developments in education policy.

The Society publishes a wide range of books, journals and databases and also maintains the largest source of chemically related information in the UK through its databanks and Library and Information Centre at Burlington House, which holds 2000 journals and in excess of 20,000 books. Computers and on-line facilities enable staff to answer a wide range of chemical queries from members and organisations via telephone, fax and email.

There are also 35 Local Sections of the Society which cater for the needs of members. Their committees often contain teacher members and frequently have links with Chemistry Teachers' Centres in their areas, providing an invaluable link between schools, colleges, industry and tertiary education. In particular, Local Sections organise schools and colleges lectures, careers talks and local competitions thereby providing a valuable local resource.

The importance of science in a broad balanced education is now fully recognised. Chemistry occupies a central position within this framework as a major subject in itself as well as a bond between physics and biology. Ironically, chemistry permeates so much of our existence that its significance can often be understated, its workings overlooked. The Society recognises that primary education is the starting point for this and so has produced this resource for teachers. The Society also produces other materials for teachers, details of which can be found on LearnNet, the new education network which provides direct access to teaching and learning resources across all age groups:
http://www.chemsoc.org/networks/LearnNet

For further information on RSC educational activities for schools and colleges, contact

Education Department
Royal Society of Chemistry
Burlington House
Piccadilly
London
W1J 0BA

Tel: 020 7437 8656

email: education@rsc.org

THE ASSOCIATION OF THE BRITISH PHARMACEUTICAL INDUSTRY

Many of the most significant medical discoveries of the 20th century were made by British scientists. Research successes such as penicillin have helped to reduce the impact of infection and disease all over the world. Former scourges such as polio and diphtheria are now rare in the developed world. Yet many diseases remain for which there are either only adequate treatments or in some cases none at all.

The Association of the British Pharmaceutical Industry is the trade association for some 80 companies in the UK producing prescription medicines. Its member companies research, develop and manufacture over 80% of the medicines and vaccines supplied to the National Health Service. Most of the medicines which your doctor prescribes have been made by its members. In developing these medicines, the industry works closely with universities and medical schools.

The pharmaceutical industry is constantly developing new medicines to treat the whole range of illness and disability. British companies have discovered and developed 6 out of the world's top 25 leading medicines and devote a huge amount of effort to research into new, safer and more effective treatments. The industry invested £2.7 billion in research and development in the UK in 1999 – more than £7 million every day.

Many industries support the education system by demonstrating the application of science and technology. Through this, they hope to excite and attract young people into science and equip pupils with the scientific knowledge and skills they will need in their adult life. The ABPI believes that primary school teachers have a key role in providing the skilled and well-educated young people who are the future scientists and technologists needed by UK companies, universities and the teaching profession itself, and in helping to improve future levels of scientific literacy.

The pharmaceutical industry has long recognised the benefits that companies and schools gain from working with each other. The industry is amongst the leaders in industry-schools partnership in terms of quality, relevance of content and in ensuring good practice, as well as its level of involvement.

The majority of benefits from schools links come from partnerships at a local community level. Organisations such as ABPI, the RSC and the Association for Science Education (ASE) aim to disseminate best practice and support education at a national level by providing information on real life applications of science and careers.

The ABPI is sponsoring "That's chemistry!" as an aid to introducing young children to chemistry and to encourage development of thinking skills through scientific investigations.

For further information on the ABPI's education activities please contact:

Education Executive
ABPI
12 Whitehall
London SW1A 2DY

email: sjones@abpi.org.uk

CONTENTS

	Page
Introduction	iii
Investigations and scientific activities	iv
Organisation	v
Differentiation	v
Tables, charts and graphs	vi
Standard units	vii
Health and safety	vii
Acknowledgements	xi
Bibliography and teachers resources	xii
Concept cartoons on materials and their properties	xxi
Partnership activities with industry	xxiii
Curriculum links on the website	xxxi
Using ICT to enhance teaching and learning	xxxiv

A Grouping and classifying materials

Grouping and classifying materials on the basis of their simple properties	1
The properties of materials and their everyday uses	11
Thermal insulation and conduction and electrical insulation and conduction	25
Rocks and soils	35
Solids, liquids and gases	53

B Changing Materials

Mixing and dissolving materials	63
Heating and cooling materials	75
Burning	87
Irreversible changes – chemical reactions	95
The water cycle – evaporation and condensation	103

C Separating mixtures of materials 115

INTRODUCTION

This book is not intended to be a scheme of work, but is a manual of ideas, activities and investigations about the science of materials for teachers to use with primary children. They may add to or offer an alternative to those already being carried out in schools. All the activities have been trialled, although in accordance with good practice should be tried by the teacher before using them with a group of children.

Each chapter begins with scientific background knowledge appropriate for the concepts covered in that section and these are intended to help the non-specialist teacher. There is a suggested vocabulary to introduce to children and a list of science skills.

Key ideas or learning intentions are followed by a variety of activities described in some detail. These are intended to help children to develop an understanding of the concepts by taking part in a practical activity themselves or watching a demonstration. Each set of activities may be targeted at a particular year group. Some ideas however, begin with activities for younger children and continue with those for older children, in order to develop the concept further. Extension ideas are often offered for more able children. The intention is for teachers to select activities that best suit their children, school and classroom circumstances.

Opportunities for the use of ICT to enhance the teaching and learning experience have been indicated where appropriate.

The activities section always includes a list of safety considerations, and in some cases additional reference is made to this within the section. It is a good idea to get into the habit discussing the hazards and risks involved with the whole class, especially for situations which maybe particularly hazardous.

Teachers should be aware of the useful guide 'Safe Use of Household and Other Chemicals' (Leaflet L5p) produced by the CLEAPSS School Science Service. However teachers should be aware that this is only available to subscribers to CLEAPSS.

Cross-curricular links

Where possible, links to other areas of the curriculum have been made and in all cases there are references to the Numeracy Strategy and the non-fiction Literacy Strategy for England and Wales. In addition, poems from the star* project have been included. This project on Science, Technology and Reading is funded by The Design Council, Esso and the Royal Society of Chemistry, with support from the Institute of Physics and the Association for Science Education. These poems have been given literacy references in number form only for the age of the children for whom the science activities are appropriate. This and other curriculum information can be found detailed on the following website; www.chemsoc/networks/LearnNet/thats-chemistry.htm

star* Project

Science, Technology and Reading is the star* project and includes a unique collection of poems especially written for it by Michael Rosen. The poems deal with specific scientific concepts in an everyday context and where appropriate, some have been included in this book. They are intended as an exciting introduction to a science lesson, offering the concept in its everyday context, a requirement of the Programme of Study for Science in the National Curriculum. In addition, they offer an opportunity to take a cross-curricular approach, by being referenced to the National Literacy Strategy, where they may also be used.

Literacy

Poetry and stories star* poems have been selected and included for all of the chapters. In addition, other poems and stories have been suggested in the resources section on pages xvii-xx, to link particular scientific concepts with the language curriculum.

Non-fiction Throughout the primary school, children are expected to experience non-fiction writing, which will often have a cross-curricular theme with a specific genre. Some suggestions have been made for this and is dependent on the suggested activities in the chapter and the age of the children carrying them out.

INVESTIGATIONS AND SCIENTIFIC ACTIVITIES

Most of the ideas presented may be carried out in a variety of ways. These include demonstrations by the teacher, investigative and explorative activities and part or whole investigations.

Practical activities such as researching a topic, illustration (eg following instructions for crystal growing), observation, surveys, learning skills (using a thermometer), and handling secondary data are not investigations. They are valuable scientific activities, which may carry specific, skills-based learning intentions. Most of the time, however, they can be incorporated into investigative and explorative activities where pupils are encouraged to make increasingly more choices and decisions for themselves. This is a valuable way of learning scientific and key skills, and building children's confidence to eventually carry out a full, independent investigation.

Investigations involve a set of scientific skills such as identifying a problem or need, planning the process and equipment, carrying out the task, observing and recording results and concluding from the results. This is time-consuming, so children should experience it when it is most appropriate, not all the time. Suggestions for investigations have been made in the book where they are particularly suitable.

Not all investigations are of the type where variables are the focus. Scientists carry out a variety of investigations where the skills of planning, carrying out the task, recording results and concluding are needed. Whilst the investigation must be carried out in a scientifically 'fair' way, changing maybe one variable, the focus maybe on another skill, depending on what is appropriate eg pattern finding. Children should experience a varied diet of investigations. The following give examples of other different investigation types:

(1) **Classifying and identifying,** eg which materials are magnetic? What is this small creature, its name and type?

(2) **Exploring.** Observations of an event over a period of time, eg germinating seeds or the development of frog spawn.

(3) **Pattern seeking.** Observing, measuring, recording, then carrying out a survey to see if there is a pattern, eg do tall people have the longest legs?

(4) **Making things or developing systems.** This maybe a combined science/technology investigation, eg designing and making a traffic light system.

(5) **Investigating models,** eg does the mass of a candle increase or decrease when it burns?

Many non-investigative activities may be turned into investigative ones if it is appropriate to do so by changing the task title. For example an illustrative activity such as making carbon dioxide using Alka-Seltzer and water can be made into an investigation by posing the question, 'Does the amount of Alka-Seltzer used affect the amount of carbon dioxide made?'

This type of activity presents opportunities for information collecting and handling using ICT.

ORGANISATION

The classroom organisation of science activities depends on a variety of factors, including the number and ages of the children in the class, the size of the room, number of adults helping, the available resources and the specific activity being carried out. Clearly one teacher and thirty 10 year old children, carrying out a 'burning materials activity' is not advisable for many reasons, not least of all the safety factor. For activities like this, it is advisable to carry them out in small, closely supervised groups where the children work in pairs. A classroom assistant or parent helper is very useful, even if not essential, for times like this.

The current pressures of curriculum delivery tend to dictate whole class teaching, which is probably what will happen for most science activities, especially with older children. Class rules about moving around, collecting, sharing and using equipment need to be established for science, as for any other practical subject. Children participate and learn best when they work in very small groups of two or three, but this could be part of a larger group of six, working at one table, who share a tray of equipment.

It is not always necessary for all the children to do all the same activities when dealing with a key idea in science. Differentiation dictates that some children will not be able to do this anyway. If, for example, children are testing materials or looking at factors affecting a scientific process such as dissolving they do not all need to do all the same tests. It is better to do two tests really well, where scientific skills are developed, than attempt to rush through more. The important factor would then be the plenary session at the end, which would bring all their ideas together in a whole class discussion and group presentation. This also addresses various aspects of the language curriculum.

DIFFERENTIATION

It is generally accepted that most children will be able to carry out a simple whole investigation independently by the time they leave primary school. Some, however, will not.

Children gradually develop the skills to do this and investigations can be differentiated not only by changing the task but also by the amount of teacher input that is given.

For example, an investigation into the factors that affect dissolving may be differentiated in many ways:

(1) Limit the number of factors that are investigated for less able children.

(2) When investigating each factor, some children may plan, choose apparatus etc for themselves, others may need help for all or some of this.

(3) You may wish to use the investigation as an opportunity to focus teach a skill to a group of children, such as improving their results chart or teaching them how to draw a line graph.

Helping in any part of an investigation does not negate its value or mean that the investigation cannot be assessed. Children may do a part investigation independently and receive help with the rest. For example, a child may need help planning an investigation, but be able to carry out the task and record results independently. It may not be possible to assess a complete investigation for some children.

(4) Some investigations are not appropriate for all the class eg investigating the saturation point of a solid in water, but may be used as an extension activity for some able 9-10 year old children.

TABLES, CHARTS AND GRAPHS

Charts and graphs help to make sense of a list of measurements and enable children to visually see the results of their scientific investigations. It also helps them to see the effect that variables have on each other. They need to develop the skills of reading and constructing charts and graphs.

There are conventions for tables, charts and graphs, all of which should be given clear, appropriate headings.

Tables should be drawn in column form. The left-hand column is for the independent variable, which is the one you choose and change, and the right-hand column is for the dependent variable, which is the one you observe and measure.

Any units used should be in the column heading not in the column.

There can sometimes be confusion about plotting variables on a graph and deciding which variable goes on which axis. The independent variable, which is chosen by the experimenter, goes on the horizontal or 'x' axis, and the dependent variable, which are the readings to be made, go on the vertical or 'y' axis.

In the science activities in this book, most of the graphs used are bar charts or line graphs.

(a) Bar charts are used when the independent variable is not numerical. The bars can be in any order, equal width and should not touch. Lines, paper strips and materials etc can be used instead of bars.

(b) Column graphs are very similar to bar charts but are used when either or both variables are whole numbers, the second variable being discreet numbers and not continuous. The columns should be in order of increasing or decreasing size.

(c) Histograms are for continuous survey data (any number) where the data has been grouped in numbers, eg weight, 0-10 kg, 10-20 kg, 20-30 kg. This is represented as a column similar to a column graph, but the columns are touching.

(d) Line graphs are also for continuous data where two variables are used eg time and temperature of dissolving. The points of the data are marked with an 'x' and the 'best fit' line or curve drawn through the points. Usually in science the points are not individually joined up, although this is common practice in other subjects. The units used are part of the axis label and not next to the data numbers. More than one line can be drawn on a graph, if the same two variables are used, to compare a third variable eg rate of evaporation of different liquids. Colours can be used to differentiate the lines.

(e) Pie charts are used for survey data as an alternative to Venn diagrams, column graphs, histograms and bar charts. There should be no more than six categories in order to avoid it becoming too complex, with the data key next to the chart. The sections should be in rank order beginning at '12 o'clock.'

Many graphs and charts can be produced using ICT packages.

STANDARD UNITS OF MEASUREMENT

These are the metre, m, the unit of length; the kilogram, kg, the unit of mass; the second, s, the unit of time; and the cubic metre, m^3, the unit of volume. As the use of litres is common in everyday life children should know both.

1 litre = 1 dm^3
1 ml = 1 cm^3
1 dm^3 = 1000 cm^3

The unit of temperature used is the degree Celsius (°C) which is **not** a standard unit.

HEALTH AND SAFETY

Any situation in school is potentially dangerous and the ultimate responsibility rests with the employer. However there is an expectation that the employee will behave in a certain way as to minimise the risks. All schools should have a school safety policy, which includes references to science activities to which teachers should adhere. In addition, most local authorities belong to CLEAPSS (Consortium of Local Education Authorities for the Provision of Science Services) and follow their guidelines. They recommend the use of the ASE booklet 'Be Safe' which offers advice for most primary science activities and are extremely helpful if additional advice is needed. (The telephone number and website is at the end of this section). However advice is only available to subscribers.

All countries in the UK make specific reference in their curriculum documents to children of all ages recognising the hazards and risks to themselves and others when taking part in a science activity. This therefore means that this awareness needs to be taught, together with strategies for dealing with specific hazards. Practical activities of any sort where children are moving about and using equipment are potentially hazardous and all such activities should be preceded by a brief discussion with children. Specific situations where risks are greater and particular strategies are used, such as the wearing of goggles will also need discussion with the children. It might also be a good idea to get them to design their own safety symbols to put in the margins of their work to highlight the need for safe working, and to use them regularly when they are needed.

HEALTH AND SAFETY

All the activities in this book can be carried out safely in schools. The hazards have been identified and any risks from them reduced to insignificant levels by the adoption of suitable control measures. However, we also think it is worth explaining the strategies that are recommended to reduce the risks in this way.

Regulations made under the Health and Safety at Work Act 1974 require a risk assessment to be carried out before hazardous chemicals are used or made, or a hazardous procedure is carried out. Risk assessment is your employers responsibility. The task of assessing risk in particular situations may well be delegated by the employer to the head teacher/science co-ordinator, who will be expected to operate within the employer's guidelines. Following guidance from the Health and Safety Executive most education employers have adopted various nationally available texts as the basis for their model risk assessments. These commonly include the following:

Safeguards in the School Laboratory, 10th edition, Association for Science Education, 1996

Topics in Safety, 2nd Edition, Association for Science Education, 1998 (new edition available in 2001)

Hazcards, CLEAPSS, 1998 (or 1995)

Laboratory Handbook, CLEAPSS, 1997

Safety in Science Education, DfEE, HMSO, 1996

Hazardous Chemicals Manual, SSERC, 1997.

If your employer has adopted one or more of these publications, you should follow the guidance given there, subject only to a need to check and consider whether minor modification is needed to deal with the special situation in your class/school. We believe that all the activities in this book are compatible with the model risk assessments listed above. However, teachers must still verify that what is proposed does confirm with any code of practice produced by their employer. You also need to consider your local circumstances. Are your students reliable? Do you have safety glasses for everyone?

Risk assessment involves answering two questions:

How likely is it that something will go wrong?

How serious would it be if it did go wrong?

How likely it is that something will go wrong depends on who is doing it and what sort of training and experience they have had. In most of the publications listed above there are suggestions as to whether an activity should be a teacher demonstration only, or could be done by students of various ages. Your employer will probably expect you to follow this guidance.

Teachers tend to think of eye protection as the main control measure to prevent injury. In fact, personal protective equipment, such as goggles or safety spectacles, is meant to protect from the unexpected. If you expect a problem, more stringent controls are needed. A range of control measures may be adopted, the following being the most common. Use:

- a less hazardous (substitute) chemical;
- as small a quantity as possible;
- as low a concentration as possible; and
- safety screens (more than one is usually needed, to protect both teacher and students).

The importance of lower concentrations is not always appreciated, but if solutions are suitably dilute they are classified as irritant rather than corrosive.

Throughout this resource, we make some reference to the need to wear eye protection. Undoubtedly, chemical splash goggles, to the European Standard EN 166 3 give the best protection but children are often reluctant to wear goggles. Safety spectacles give less protection, but may be adequate if nothing which is classed as corrosive or toxic is in use. It is recommended that corrosive or toxic materials are **not** used in primary schools.

CLEAPSS

Teachers should note the following points about CLEAPSS:

At the time of writing, every LEA in England, Wales and Northern Ireland (except Middlesbrough) is a member, hence all their schools are members, as are the vast majority of independent schools, incorporated colleges and teacher training establishments and overseas establishments.

Members should already have copies of CLEAPSS guidance in their schools.

Members who cannot find their materials and non-members interested in joining should contact the CLEAPSS School Science Service at Brunel University, Uxbridge, UB8 3PH. Tel: 01895 251496, fax: 01895 814372, email: science@cleapss.org.uk or visit the website http://www.cleapss.org.uk.

Schools in Scotland have a similar organisation, SSERC (Scottish Schools Equipment Research Centre), 2nd Floor, St Mary's Building, 23 Hollyrood Road, Edinburgh EH8 8AE. Tel: 0131 558 8180.

Chemicals used in this book

Chemical	Hazard
Aluminium potassium sulfate	None
Borax (sodium tetraborate decahydrate)	None
Dilute hydrochloric acid 0.5 mol dm^{-3}	None
Iron(III) oxide	None
Iron filings	None
Plaster of Paris (anhydrous calcium sulfate)	None
Poly Vinyl Acetate (PVA) glue	None
Sodium hydrogen carbonate	None
Universal Indicator solution	None

HEALTH AND SAFETY

Hazard and risk in teaching

Discuss with children the hazards of an activity – what might go wrong? Then discuss the risks – how likely is it that something would go wrong? How many people could be affected? How badly would each be affected? (One hand chopped off would be a big problem; a minor cut on a hand would be a small problem; a class set of minor cuts on hands would be a medium problem). Then discuss the control measures – how can we make it less likely that something does go wrong?

The word 'danger' can be used when you want to talk generally, but to be precise then hazard, risk, etc are the correct terminology.

So, as an example, for elastic bands, the hazards might be :

(a)   Breaking, thus dropping a heavy load on your toe;

(b)   Breaking, contracting and flicking something into your face or eye;

(c)   Being misused to fire pellets at each other.

The chance of (a) actually happening may be zero (if you are not using heavy loads). Or considerable if you are! The number of people affected may depend on whether the whole class is doing this or just one group. How serious the injury might be would depend on the type of shoes being worn, how heavy your load really is, etc. Whether it will drop onto toes or not may depend on whether there is a bin underneath to keep toes out of the way (a control measure). The chance of (c) happening will depend on many factors such as the control exercised by the teacher and the normal behaviour of the class.

ACKNOWLEDGEMENTS

The author would like to thank a number of people for their help in producing this book.

Colin Osborne and the team in the Education Department of the Royal Society of Chemistry, Burlington House, London, for their support and guidance.

Mike Willson and the PGCE Science Team at Sussex University for letting her share their space, knowledge and jokes!

John Reynolds, ESTA and John Payne for their assistance with Earth Sciences.

Peter Borrows, CLEAPSS for answering numerous questions about safety.

Brighton University, for the use of the Education Department library facilities.

Geoff Hockney, University of Sussex, for developing my computer skills.

Doug Dickinson, for assistance with ICT.

The following people trialled star* poetry and the science activities or allowed the author to do it with their children:

Viv Aylwood (PGCE student, Sussex University)	Somerhill Junior School, Hove.
Mary Hinton	
Lesley McGorrigan	
Wenda Bradley, Headteacher.	
Rose Watkin, Deputy Headteacher and all the staff	Sidlesham County Primary, Chichester.
Franca Reid, Headteacher and staff.	Longforgan Primary School, Scotland
Dr. Stephen Turner, RSC Huddersfield Local Section	Fixby Junior and Infant School. Christ Church Woodhouse, C of E School
	Bradley C of E School.
Lisa Smith	Wood End Junior School, Greenford, Middx.
David Brown, Headteacher and Staff	Lodge Mount Primary School, Loughborough.

Staff and children at St. Paulinus C of E Primary School for cover photos.

The Royal Society of Chemistry would like to thank Professor Joan Bliss and the University of Sussex Institute of Education for providing office accommodation, and awarding Jan Rees the honorary title of Visiting Research Fellow from September 1 1998 to August 31 1999.

BIBLIOGRAPHY
and teachers resources

D. Archer, *What's Your Reaction,* London: Royal Society of Chemistry, 1991.

ASE *Be Safe!,* Hatfield, Herts: Association for Science Education, 1990 (Scottish edition 1995).

ASE *Signs and Symbols in Primary Science,* Hatfield, Herts: Association for Science Education, 1998.

P. Bell, *National Curriculum* Theory into Practice, Materials and Change, (Topical Resources, 1993).

T. Deary, and B. Allen, *The Spark Files, Chop and Change,* Faber and Faber, 1998.

J. Davy, *Memoirs of the Life of Sir Humphry Davy,* Vol 2, Longman, 1836.

Teaching Primary Earth Science, Earth Science Teachers Association, ESTA 1994-9.

R. Feasey, *Primary Science and Literacy Links,* Hatfield, Herts: Association for Science Education, 1999.

R. Frost, *IT in primary science,* Hatfield, Herts: Association for Science Education, 1999.

A. Goldsworthy, and R. Feasey, *Making Sense of Primary Science Investigations,* Hatfield, Herts: Association for Science Education, 1997.

P. Gannon, *KS3 Science The Revision Guide,* Sc. Co-ordinators Group. Kirby-in-Furness. Cumbria, 1998.

J. Hann, *How Science Works,* Eyewitness Science Guides, Dorling Kindersley, 1991.

R. Heddle, *Science in the Kitchen,* Usborne, 1992.

M. Hoath, *100 Ideas for Science. 7-11,* London: Collins Educational, 1998.

M. Hollins, *Materials,* BBC fact Finders, BBC Educational Publishing, 1994.

L. Howe, *Collins Primary Science KS1 and KS2, Things Change; Solids, Liquids, Gases; Over, Under and all Around; Nursery Rhymes; Stories; Eggs; Water; Clothes; Drinks; Wet and Dry; Colour; Our Senses,* Harper Collins, 1990.

T. Jennings, *The Young Scientist Investigates,* Book 2, Oxford: Oxford University Press, 1986.

B. Keogh, and S. Naylor, *Concept Cartoons,* Millgate House, 1997.

S. Naylor and B. Keogh *Concept Cartoons in Science Education,* Millgate House, 2000.

Learning Through Science Scheme, Materials and Colour, Macdonald, 1985.

Learning Through Science Scheme, Earth, Macdonald, 1985.

G. Matthews, *100 Ideas for Literacy, Non-fiction,* London: Collins Educational, 1998.

SC1 Investigations. KS1/2, Hatfield, Herts: Association for Science Education, 1998.

Nuffield Combined Science. *Middle years,* Clothes. Fabrics, 1970.

Nuffield Primary Science. *Science and Literacy,* London: Collins Educational, 1998.

Nuffield Primary Science. *Rocks and Soils*, 5-7, 7-12. Teacher's Guide, London: Collins Educational, 1993.

Nuffield Primary Science, *Materials*, 5-7, 7-12. Teacher's Guide. London: Collins Educational, 1993.

New Horizons. 5-16 KS1/2 *Stories to illustrate Science Concepts*, Cambridge: Cambridge University Press, 1992.

Oxford Science Programme KS3. *The Earth in Balance*, Oxford: Oxford University Press, 1991.

Oxford Science Programme KS4 *The Earth and Beyond*, Oxford: Oxford University Press, 1991.

S. Parker, *Fun with Science, Simple Chemistry*, Kingfisher, 1990.

J. Parvin, *Kitchen Concoctions: A Pinch of Salt, Hats off to Dora, 8-11 years, Water for Industry, 10-12 years, Tidy and Sort, 5-7 years, Pencils, Poems and Princesses*, University of York, Chemical Industry Education Centre, 1994.

M. Revell, and G. Wilson, *Chemistry and Cookery*, Northamptonshire Science Resources, 1995.

M. Revell, and G. Wilson, *Storybooks and Science*, Northamptonshire Science Resources, 1995.

M. Revell, and G. Wilson, *Science from Stories*, Northamptonshire Science Resources, 1995.

SATIS 8-14*, Box 1 and 2, Hatfield, Herts: Association for Science Education publications, 1992.

Using Stories to Stimulate Science and Technology With the Under 5's, Sheffield Education Authority, 1993.

M. Stephenson, and L. Hubbard, *Primary Good Resource Guide*, University of York, Chemical Industry Education Centre, 1997.

J. Tavener, *Counting Rhyme Activities*, Scholastic, 1997.

A. Treneer, *The Mercurial Chemist*, Metheun, 1963.

B. Walpole, *Salt*, A. C. Black, 1991.

W. Wade, and C. Hughes, *How to be brilliant at Materials*, Brilliant publications, 1997.

M. Wenham, *Understanding Primary Science Ideas and Concepts*, Paul Chapman, 1995.

The National Curriculum for Science, DfEE and QCA, 1999.

The National Literacy Strategy, DfEE and QCA, 1998.

The National Numeracy Strategy, DfEE and QCA, 1999.

Ginn Science and Star Science Scheme. Materials, Ginn, 1988.

BIBLIOGRAPHY

Other Resources

Three Valleys Water, The Water Box, Three Valleys Water. Tel: 0345 724665

Earth Science Teachers Association (ESTA) Publications rock samples, and activities. For information contact: Mr P. York, 346, Middlewood Road North, Outibridge, Sheffield, S35 0HF.

Candle Makers Supplies Ltd. 28, Blythe Road, London. Tel: 020 7602 4031

Further information on materials

There are many publications on different types of materials, which can be borrowed from the library or purchased relatively cheaply. These range from very easy reading for younger children to more complex text.

Your library may also offer CD-ROMs and good information can also be found in a variety of encyclopaedias.

Visits

Museums

The Science Museum

Exhibition Road
South Kensington
London SW7 2DD
Tel: 020 7942 4455

Website: http://www.nmsi.ac.uk (accessed 08/11/00)

The museum has an excellent gallery all about materials as well as other interactive galleries covering all aspects of science.

The Natural History Museum

Cromwell Road.
London SW7 5BD
Tel: 020 7942 5011

Website: http://www.nhm.ac.uk (accessed 08/11/00)

The museum has two galleries displaying natural materials, The Mineral gallery and a new Earth's Treasury gallery, which is interactive.

The Victoria and Albert Museum

At the corner of Cromwell Road and Exhibition Road,
South Kensington
London
Tel: 020 7942 2680

Website: http://www.vam.ac.uk (accessed 08/11/00)

This museum has two study galleries about materials, one displays different textiles, the other is about different materials used for ornamental decoration.

Local Museums

Many small museums around the country display examples of local rock and rock formation.

Local Industry

Local industries often have open days where the public can see around the works. More formal partnership activities with industry are described on page xxi.

Local Water Companies

Many local water companies produce leaflets about the 'Water Cycle' and also have 'Open Days' for the public. To find out about your local company you can contact the Water Services Association Tel: 020 7957 4567

Three Valleys Environmental Centre

Clay Lane Treatment Works.
Clay Lane. Bushey, Herts, WD2 3RE
Tel: 0345 724665

This is a local water company which offers a centre for parents and children to visit.

email: tcasey@3valleys.co.uk

Websites

The Victoria and Albert Museum: http://www.vam.ac.uk (accessed 08/11/00)

The Natural History Museum: http://www.nhm.ac.uk (accessed 08/11/00)

The Science Museum: http://www.nmsi.ac.uk (accessed 08/11/00)

Water Services Association: http://www.water.org.uk (accessed 08/11/00)

Other useful websites can be accessed via
http://www.chemsoc.org/networks/LearnNet/thats-chemistry.htm

Organisations

The Royal Society of Chemistry
Burlington House
Piccadilly
London
W1J 0BA

Tel: 020 7437 8656

http://www.chemsoc.org (accessed 08/11/00)
http://www.rsc.org (accessed 08/11/00)
http://www.chemsoc.org/networks/LearnNet (accessed 08/11/00)

BIBLIOGRAPHY

Earth Science Teachers Association.
Mr P. York.
346, Middlewood Road North
Outibridge
Sheffield
S35 0HF
Publications, activities and rock samples

The Association of the British Pharmaceutical Industry
12 Whitehall
London
SW1A 2DY

Tel: 020 7930 3477

http://www.abpi.org.uk (accessed 08/11/00)

BIBLIOGRAPHY

Suggested Poetry for Materials and their Properties

A Picnic Of Poetry	Anne Harvey	Puffin
Baking Day	Rosemary Joseph	
Left-handed Egg	John Pudney	
Burnt Carrots	Michelene Wander	
Egg Thoughts	Russell Hoban	
On Making Tea	R. L. Wilson	
A Second Poetry Book		Oxford University Press
The fate of an icicle	Alan Sillitoe	
The Toaster	William Jay Smith	
A Fourth Poetry Book		Oxford University Press
The Building Site	Gareth Owen	
Another Fourth Poetry Book		Oxford University Press
The Touch of Sense	John Kitching	
Ice on the Rond Pond	Paul Dehn	
A Word of Poetry (KS2)	Selected by Michael Rosen	Kingfisher
The Fire of London	Jim Wong Chu	
Water		
Ice		
Comic Verse	Selected by Roger McGough	Kingfisher
Death of a Snowman	Vernon Scanell	
I'd Like to be a Teabag	Peter Dixon	
Kenneth, who was too fond of bubblegum and met an untimely end	Wendy Cope	
Balloon	Colleen Thibandean	
First Poems (KS1/2)	Julia Eccleshare	Orchard Books
Mary and Sarah	Richard Edwards	
The Snowman	Roger McGough	
The Jumblies	Edward Lear	
Food Poetry	Robert Hull	Wayland
On making Tea	R. L. Wilson	
Blue Peter	Mick Gower	
I Didn't Want To Come To Your Party Anyway	Rita Ray	
Oxford Treasury of Children's Poems	Michael Harrison, Christopher Stuart-Clarke	Oxford University Press

BIBLIOGRAPHY

Egg Thoughts	Russell Hoban	
The Toaster	William Jay Smith	
Boiling An Egg	Stanley Cook	
The Snowman	Roger Mc Gough	
Ice	Walter De La Mare	
Please Mrs Butler	Allan Ahlberge	Puffin
Only Snow		
Balls on the roof		
Poems about Weather	A. Earl and D. Sensier	Wayland
I Wonder	Opal Patmer Adisa	
Winter Morning	Frank Flynn	
Cloudburst	Richard Edwards	
Shower	Moira Andrew	
Scrumdiddy	Jennifer Curry	Red Fox
Making Chocolate Cake	Janis Priestley	
Egg Thoughts	Russell Hoban	
Science Poetry	Selected by Robert Hull	Wayland
Electricity	Rebecca Hughes	
Facts About Air	John Foster	
Crystals	Barrie Wade	
Bubbles	Ryan Goad	
Sense and Nonsense	Susanne and Shona Mckellar	Macdonald
Tasting		
Listening		
Touching		
Looking		
The Word Party	Richard Edwards	A Young Puffin
Cloudburst		
When We Were Very Young	A. A. Milne	Metheun
Happiness		
Sand – between – the toes		
Time for One More	Leila Berge	Metheun
Wellingtons		
Whispers From a wardrobe	Richard Edwards	Puffin
Tears on Monday		
Whistling		
The Rain		
Walking The Bridge of Your nose	Word play poems selected by Michael Rosen	Kingfisher
Esau Wood		
Down the slippery slide		
Scouring round the porridge pot		

Wondercrump Poetry	Roald Dahl Foundation, By children for children	Red Fox
Bubbles	Aaron Davies	
The Rain Poem	Vicky Slight	
Rain	Daniel Williamson	
Fog	Fahad Shahid Sayood	
Vesuvius – Tears of Fire	Almeena Ahmed	

Nursery Rhymes

Incy, Wincy Spider	Puddles-evaporation
Jackfrost	Change of state, water, ice
Doctor Foster	Rain, watercycle
The Queen of Hearts	Cooking, effect of heat on pastry and jam
Polly Put the Kettle on	Evaporation and condensation
One Misty, Moisty Morning	Solids, liquids gases, water vapour
London Bridge is Falling Down	Strength of materials, metals rusting
Pat-a-cake Baker's Man	Cooking, effect of heat and mixtures
Five Currant Buns	Cooking, effect of heat and mixtures
Baa, Baa, Black Sheep	Types of fabrics-wool. Strength of bags to hold the wool
Old King Cole	Pipe, bowl, materials for different uses

Traditional Stories for Material and their Properties

Sleeping Beauty	Changing materials over time, erosion, rusting, weathering, rotting
Goldilocks	Strength of the materials used for beds and chairs Temperature of the porridge and how to keep it warm. Which size bowl cooled the quickest?
The Three Little Pigs	The best materials for building the house, a waterproof roof
Dick Whittington and *Puss in Boots*	Materials for the strongest boots Materials to make a strong, waterproof cloth to carry things in on his stick
Babes in the Wood	Keeping warm in the woods at night – insulation. Make a 'nest' which would keep things the warmest

BIBLIOGRAPHY

Little Red Riding Hood	Insulation. Keeping food warm on the way to grandma's house
Rapunzel	How strong is hair? Test hair and other threads for strength
Hansel and Gretel	Cooking. Make an edible house, gingerbread mixture, icing sugar
The Gingerbread boy	Cooking. Making gingerbread, using different recipes
The Emperor's New Clothes	Transparent, opaque, translucent, different types of fabrics
The Steadfast Tin Soldier	Effect of heat on metals
The Wizard of Oz	Tin man – lubricating the joints, rusting. Metals-are they all magnetic?
The Remarkable Rocket (Oscar Wilde)	Burning
The Happy Prince (Oscar Wilde)	Precious rocks, effect of heat on materials
Daedalus and Icarus (Greek Myth)	Effect of heat on materials
Why the Sea is Salt (Norwegian tale) Faber Book of Folk Tales (Faber, 1980)	Dissolving
How Rabbit Stole the Fire. (American Indian Tale) Joanna Troughton, (Blackie, 1983)	Burning

CONCEPT CARTOONS ON MATERIALS
and their properties

Concept cartoons are drawings that illustrate possible areas of uncertainty in everyday situations. The six cartoons reproduced in this book are reproduced with permission from 'Starting Points for Science' by Brenda Keogh and Stuart Naylor, Millgate House Publishers, 1997.

Each cartoon presents children with alternative viewpoints on some scientific concept. They are thus a useful starting point for both discussion and investigation in order to explore which of the alternatives are likely to be correct.

Concept cartoons are not necessarily designed to have a single right answer. In many cases the only possible answer is 'It depends on…' This is a realistic perspective on science for children to develop and can dispel the myth that there is always a 'right' answer. A variety of investigative approaches can be developed from the cartoons depending on the ideas produced by the children.

Teaching strategies can include group and class discussion or as starting points for individual projects or as an extra challenge to able pupils. However the cartoons are also useful in providing starting points for those with poor literacy skills, those who lack confidence in science, and those who are reluctant learners.

The particular cartoons included are described below.

Snowman (page 32)
The issue in this concept cartoon is whether the coat is an insulator or whether it actually generates heat. Some children may believe that warm clothes make you warmer by making more heat, and they will expect the coat to generate heat and melt the snowman faster. However others will realise that the coat is simply an insulator which will tend to keep heat away from the snowman and prevent it from melting quickly. The situation shown in the concept cartoon can be investigated using real snow. Alternatively it can be modelled with ice inside a coat, glove or sock; the top half of a plastic mineral water bottle, filled with water and frozen, will make a good model snowman. The thickness, colour and nature of the material that the coat is made from can also be investigated.

Ice pops (page 31)
All of the predictions in this concept cartoon can be directly investigated by the children. Some of them are likely to think that aluminium foil is an insulator; that cotton wool makes things warmer; that water will keep the ice pop cold; and that things will stay frozen inside a refrigerator. In each case they will be surprised by their observations! This can lead on to a whole series of follow up investigations on conductors, insulators and heat transfer.

Ice cream (page 112)
Although the children will have experience of condensation they are unlikely to have well-formed ideas about where the condensed water comes from. The concept cartoon invites them to consider and investigate a number of possibilities, and they may well think of other possibilities themselves. The fact that the condensation comes from the air may appear to be the least likely possibility to many of the children. Wrapping the ice cream tub in polythene or aluminium foil and observing where the condensation forms should help to clarify their ideas. Investigations such as this help to lay the foundation for later work on the structure of matter and conservation of mass.

Is it a solid? (page 61)	The children will have intuitive ideas about what they mean by a solid. However they will not find it easy to apply their ideas in a consistent way. They will find it difficult to separate the object from the material it is made from, and they will tend to associate properties such as heaviness and rigidity with solids. The concept cartoon provides an opportunity for them to rethink their definitions and to make more systematic judgements. Introducing more challenging materials such as sand or dough is probably best left until after their ideas about solids are reasonably well developed.
Sandcastles (page 50)	The distinction between melting and dissolving is a common area of confusion for children. They can clarify the meaning they attach to both of these terms by investigating the situation shown in the concept cartoon. A tray full of sand can be used to model the effect of the tide on sandcastles. Observation of other changes in materials, such as melting chocolate or dissolving sugar, will be a useful complement to their investigation.
Sugar in tea (page 123)	This concept cartoon invites the children to reverse the familiar process of dissolving. It also challenges the children's ideas about what happens to the sugar in the tea – does it disappear completely as it dissolves or can it be recovered from the tea? The children can investigate the possibilities shown in the concept carton as well as other possibilities that they might suggest. Salt is a useful alternative, since it can be separated more easily from water than the sugar. Other means of separating materials would be useful ways to follow up this investigation.

The ConCISE project can provide further details and more examples of concept cartoons. In England and Wales local authority advisers will be aware of the project and many will have offered INSET to teachers. Advisers in Scotland and Northern Ireland are being contacted in the autumn of 2000.

A further selection of Concept Cartoons are to be found in *Concept Cartoons in Science Education* by Stuart Naylor and Brenda Keogh, Millgate House Publishers, 2000.

PARTNERSHIP ACTIVITIES WITH INDUSTRY

Industry is a valuable resource to enhance not only science teaching but a host of other activities in the primary curriculum. However many teachers are unaware of the opportunities offered and the ways to go about the liaison. The information below shows you how to either make your own approach to a company or to find out about schemes such as the Royal Society of Chemistry's Chemistry at Work scheme. Details of a Chemistry at Work activity are given on page xxix.

HOW TO ORGANISE AN INDUSTRY LINK

How to get started Obviously you will not be the first to attempt this, but the company you approach may not have been involved with primary schools before, so the following suggested steps may help you and the company work out what you can do together successfully.

Step 1: How do I find the right company to approach – close ties with a local company may make things easier, so don't forget to use the children as the first point of contact. Parents, relatives, close family friends and neighbours may turn out to work in industry and the personal contact is invaluable. Failing that, school staff and governors should be the next avenue to pursue.

Your Local Education Authority can put you in touch with your regional Education Business Partnership, SETPOINT or SATRO (if they exist). You can also find details on the SETNET website (http://www.setnet.org.uk)

Suitable companies to approach to help support science teaching

The types of company you may wish to consider are:

(a) Heavy chemical manufacture large tonnages of chemicals usually extracted from oil.

(b) Fine chemical manufacture unfamiliar chemicals which are used to make more complex chemicals eg pharmaceuticals.

(c) Pharmaceutical and Health care

(d) Other processing industries manufacturers of cement, bricks and other construction materials

(e) Food processors breweries, dairies, bakeries

(f) Cosmetics

(g) Glass industry

(h) Metal foundries

(i) Plastic processors

(j) Quarrying and mining

(k) Environmental monitoring and analysis
   water treatment
   waste disposal
   recycling plants
   public health laboratories
   sewage treatment
   forensic laboratories
   hospital pharmacies

PARTNERSHIP ACTIVITIES WITH INDUSTRY

Step 2: How should I make the first contact? If you have been able to use a personal contact to establish a link with a company you will find it easier to make the first approach as you will have a name, and perhaps a bit of background information about this person and the company. This will always help to get things going.

If you don't know the name of the person to contact, it is a good idea to telephone the switchboard and ask for the name of the senior person in one of the following departments:

Human Resources (personnel)

Community Affairs (public relations)

If you have the impression that the company is small, just ask for the name of the chief executive on site.

Once you have the name of a person to contact:

(a) Write to say you will be telephoning to discuss the possibility of working with the company to help promote science in your school. (Draft letter below.)

> Dear
>
> I am writing to enquire whether there would be an opportunity for me to meet you to discuss the possibility of working with your company to enhance science teaching in my school.
>
> There are a number of activities which have been successful for other school industry partnerships. I am enclosing a photocopy of some quotations from teachers and industrialists who recommend such activities (enclose a photocopy of the page of quotes).
>
> I am interested in (choose one or two from the list) site visits/project work/a speaker coming to talk about an aspect of his work which fits the curriculum/help with science clubs/help with science investigations/ professional development courses/teacher placements – but would be happy to discuss other ideas too.
>
> Unless I hear from you I will telephone to arrange a date and time to meet.
>
> Yours sincerely

(b) Telephone and arrange a date, time and venue (either at the site or school) for a meeting.

(c) Meet with an open agenda (see step 3) but with some ideas of what you want to do (see step 4).

Step 3: Create an agenda for the first meeting Think about what you want from the meeting. Don't be too ambitious in the first instance, and allow for the relationship to develop.

Example agenda:

(1) Teaching science in primary schools
- what is in the curriculum
- how it's resourced
- the needs of the teachers

(2) The company's experience of working with primary schools

(3) List activities which are feasible for the company and desirable for schools

(4) Choose one activity to follow up
- list actions to be taken in advance
- propose a timetable
- allocate responsibilities (including costs)

Step 4: The first meeting Be open-minded, as the company, inspired by your interest, may suggest activities which you had not thought of but which are spot on. However it is likely that the company, especially if new to this kind of partnership, will want to hear your ideas first. Don't be too ambitious, but give the company a list of activities which will help enhance science teaching in your school (see letter and page of quotes).

Make sure you leave the meeting with a clear understanding and agreement of what the next action is. It's a good idea to send the company a copy of any notes taken, especially those related to the proposed activity.

You and the company will probably find yourselves very busy preparing the agreed activity, but once you see the children's reaction you will know it's been worthwhile. Parents will probably be interested to hear about the event, so a letter home telling them about it is a good idea.

Step 5: The agreed action As the event takes place don't forget you can use it to win good publicity for the school and company.

Step 6: After the event Give the company plenty of feedback, evaluating the event and if necessary suggesting improvements, but don't forget to thank them and if things have worked well suggest a meeting date to discuss the 'next' event!

PARTNERSHIP ACTIVITIES WITH INDUSTRY

FURTHER HELP	Organisation	Help offered
	Association for Science Education, ASE College Lane, Hatfield, Herts, AL10 9AA Tel: 01707 283000 Fax: 01707 266532 email: ase@asehq.telme.com	ASE membership offers: Magazines, Curriculum guidance. Safety advice, Mail order service for affordable books, National and regional events and inservice training.
	Association of the British Pharmaceutical Industry, ABPI 12 Whitehall, London, SW1A 2DY Tel: 020 7930 3477 Fax: 020 7747 1413 email: sjones@abpi.org.uk	Contact the Education Executive for information on the ways in which the ABPI and its member companies support education in the UK.
	Institution of Chemical Engineers Davis Building, 165-189 Railway Terrace, Rugby, Warwickshire, CV21 3HQ Tel: 01788 578214 Fax: 01788 560833 email: profdev@icheme.org.uk	IChemE has sponsored four self-contained science activities boxes targeted at Key Stage One and key Stage Two. For further information on the boxes contact the Professional Development Area.
	The Chemical Industries Association working through Chemical Industry Education Centre, Department of Chemistry, University of York, Heslington, York, YO1 5DD Tel: 01904 432523 Fax: 01904 434078 email: ciec@york.ac.uk	For help on working with local companies, or for information about science and technology teaching and learning.
	The Royal Society of Chemistry Burlington House, Piccadilly, London, W1J 0BA Tel: 020 7437 8656 Fax: 020 7287 9825 email: education@rsc.org	The RSC produces a wide range of careers material including leaflets and posters which are suitable for use with older primary pupils and are available free of charge. The RSC also produces curriculum material and organises 'Chemistry at Work'.

THE BENEFITS OF WORKING WITH A LOCAL COMPANY

"Teacher placements into industry are an ideal vehicle for industrialists and educationalists to increase their mutual knowledge and understanding and provide a valuable opportunity to develop long-term links and partnership activities."
Jane Gamble, ICI Teesside Education and Ecology Manager, Middlesborough.

"We have a stake in educating youngsters… so we need to become personally involved. We need to become more knowledgeable about practices in education and about barriers to better learning so we know how best to use our resources. To be effective, we need to be accepted by those within the educational community as partners, not as business people trying to tell teachers how to do their jobs."
Colin Coates, Manager of Environment, Health and Safety, Searle, Morpeth.

"Along with a group of primary teachers, not specially trained in science teaching, I went to an INSET event at Exchem Organics. There we met a group of keen young industrialists. Together we worked on a range of teaching and learning resources sponsored by industry. The outcome was an enthusiastic exchange of ideas which enhanced our motivation to teach children science with a whole new approach based on industrial contexts."
Sheila Braithwaite, Perryfields County Infant School, Chelmsford.

"Industry can provide a focus for much of the curriculum teachers deliver in the classroom. This gives an opportunity for the work to be placed in context in a more lively and interesting way."
John Adams, Academic Liaison Manager, Pfizer, Kent.

"We had scientists come into school on a number of occasions to work with the children. The children not only enjoyed all the experiments but found the opportunity of working with 'real' scientists very rewarding. Later the children visited the laboratories where the scientists worked and discovered to their amazement and delight that the experiments they were doing in school were actually used in the real world in real laboratories."
Mavis Hardwick, Hartburn Primary School, Stockton-on-Tees.

"I really enjoyed the whole experience (developing teaching resources associated with site visits) and considered that there are many advantages, such as preparing my pupils for life in a technology based society, helping to change stereotypical images which research shows still exist and raising my children's awareness of the nature of industry and its role in society."
Catherine Sinclair, McLean Primary School, Fife.

"Linking with industry delivers real learning value to the curriculum. Children see that the science work they study is valid, relevant and linked to the real world. Excellent projects within the links have given inspiration to both sides."
Barbara Pollard, Hawes Down Infant School, West Wickham.

"I never imagined that being in the classroom could be so exhilarating. Our company wanted to help create a sense of excitement during primary school science activities, something we called the 'cor' factor, but it was me saying 'cor.' I didn't know children could study so effectively and be so immersed in their work. I was also fascinated by their views on my work place and delighted that we could show them what it's really like."
Dai Hayward, General Manager, Thomas Swan and Co., Consett.

"When my class visited a local pharmaceutical company they were fascinated to see the bottling plant in action. This led to a useful project on friction and forces. Other valuable activities included a 'Health and Safety – spot the sign' trail and visits to various interactive exhibitions."
Jill Matthews, Upton County Junior School, Broadstairs.

"…it is important that industry creates a good interface with the local community and educational establishments. Being involved in science events is one of the many ways both sides benefit."
Bob Tomlin, Personnel Development Officer, BASF, Seal Sands.

"The ecological environment placement on Teesside greatly increased my knowledge… there is a wealth of knowledge, particularly scientific, to be tapped."
Pauline Bennett, St Clare's RC Primary School, Middlesborough.

This information is reproduced with grateful thanks to the Chemical Industry Education Centre at the University of York.

CHEMISTRY AT WORK – The Legoland primary event

The Royal Society of Chemistry runs a series of events called Chemistry at Work whose aims are

- to show young people how chemistry principles are applied in industry, research and everyday life;
- to show a positive image of chemistry as a rewarding, interesting and wealth-creating activity;
- to encourage students to consider careers in chemistry; and
- to show students the importance of chemistry to society.

Events take place over one to three days and groups of students spend half a day at the event visiting about half a dozen twenty-minute-long, hands-on presentations by local companies (or other organisations that use chemistry in their work). Just a few examples of presentations at recent events include:

- dry cleaning, by a consultant dry cleaner;
- testing for diabetes, by a local hospital;
- drug design by SmithKline Beecham;
- platinum in car catalysts, by Johnson Matthey;
- soft drink manufacture, by Coca Cola; and
- the chemistry of cosmetics, by a local college.

There are currently some 15 events each year around the country and the number is rising.

Most of the events are aimed at secondary school students but for the second year running, a primary school event has been run very successfully in the Bracknell area, organised in conjunction with the South East Berkshire Education Business Partnership. This particular event was hosted by Legoland Windsor which provided a splendid venue and most of the schools took the opportunity to extend their visit to a full day with the other half of the day visiting Legoland itself (free of charge!). The chemistry activities at the event included:

- a demonstration (with flashes and bangs!);
- kitchen sink science;
- the chemistry of eggs;
- the chemistry of sport;
- genetic fingerprinting; and
- aromatherapy among many others.

The event was supported by a range of organisations including The Army, Zeneca Agrochemicals, Safeway and Reading University. Feedback suggested that children (and their teachers) had an enjoyable and stimulating day and it is hoped that the event will become an annual one.

There are obvious benefits to the host and to The Society in this sort of arrangement and we will be exploring the possibilities of events at other similar venues for the future.

For further details contact

Ted Lister
Chemistry at Work National Co-ordinator
93A Upper Holly Walk
Leamington Spa
CV32 4JS
Tel/Fax: 01926 420766
email: ted@tedlis.demon.co.uk

CURRICULUM LINKS ON THE WEBSITE
http://www.chemsoc.org/networks/LearnNet/thats-chemistry.htm

The material below is included in the website with up to date curriculum links.

This book is not intended to be a scheme of work, but is a manual of ideas, activities and investigations about the science of materials for teachers to use with primary children. They may add to or offer an alternative to those already being carried out in schools. All the activities have been trialled, although in accordance with good practice should be tried by the teacher before using them with a group of children.

The book systematically covers the concepts dealt with in the Year 2000 revised Science Curriculum, Attainment Target 3, Materials and their Properties for England and Wales. It also is referenced where appropriate to the Science Curriculum for Northern Ireland and Environmental Studies 5-14 Curriculum for Scotland. It covers the chemistry, and some of the physics, studied in primary schools.

Each chapter begins with scientific background knowledge appropriate for the concepts covered in that section and these are intended to help the non-specialist teacher. There is a suggested vocabulary to introduce to children and a list of science skills.

Key ideas or learning intentions are followed by a variety of activities described in some detail. These are intended to help children to develop an understanding of the concepts by taking part in a practical activity themselves or watching a demonstration. Each set of activities may be targeted at a particular year group according to the Statements of Attainment in the curriculum. Some ideas however, begin with activities for younger children and continue with those for older children, in order to develop the concept further. Extension ideas are often offered for more able children. The intention is for teachers to select activities that best suit their children, school and classroom circumstances.

The activities' section always ends with a list of safety considerations, and in some cases additional reference is made to this within the section. It is a good idea to get into the habit of discussing hazards and risks with the whole class, especially for situations which may be particularly hazardous. This is also a statement in Science Attainment Target 1, Scientific Enquiry, of the National Curriculum.

LITERACY

Poetry and stories star* poems have been selected and included for all of the chapters. In addition, other poems and stories have been suggested in the resources section at the end of this book, to link particular scientific concepts with the language curriculum. The star* poems have been given specific Literacy Strategy links in this number form, the YEAR, TERM and then the STATEMENT from the strategy. So that 1. 3. 4. refers to year 1, term 3, statement 4. There are also very general statements from the strategy that are common to many poems which have not been mentioned specifically eg performing poetry, collections by a specific author.

Non-fiction Throughout the primary school, children are expected to experience non-fiction writing, which will often have a cross-curricular theme with a specific genre. Some suggestions have been made for these and are dependent on the suggested activities in the chapter and the age of the children carrying them out. For example, year 4 term 3 requires children to design an advertisement for a product. This might follow a science activity on dissolving and jelly making.

INVESTIGATIONS AND SCIENTIFIC ACTIVITIES

Most of the ideas presented may be carried out in a variety of ways. These include demonstrations by the teacher, investigative and explorative activities and part or whole investigations.

Practical activities such as researching a topic, illustration (eg following instructions for crystal growing), observation, surveys, learning skills (using a thermometer), and handling secondary data are not investigations. They are valuable scientific activities, which may carry specific, skills-based learning intentions. Most of the time however, they can be incorporated into investigative and explorative activities where pupils are encouraged to make increasingly more choices and decisions for themselves. This is a valuable way of learning scientific and key skills, and building children's confidence to eventually carry out a full, independent investigation. Investigations involve a set of scientific skills such as identifying a problem or need, planning the process and equipment, carrying out the task, observing and recording results and concluding from the results. This is time-consuming, so children should experience it when it is most appropriate, not all the time. Suggestions for investigations have been made in the book where they are particularly suitable.

In the revised orders for the Science Curriculum 2000 in England and Wales, there is a reference in Attainment target 1 that across Key Stage 1 and 2 children should experience different types of investigations. This is because the interpretations of the previous curriculum have led to children often only experiencing 'fair testing' type investigations such as 'How does temperature affect dissolving'?

Not all investigations are this type where variables are the focus. Scientists carry out a variety of investigations where the skills of planning, carrying out the task, recording results and concluding are needed. Whilst the investigation must be carried out in a scientifically 'fair' way, changing maybe one variable, the focus maybe on another skill, depending on what is appropriate eg pattern finding. Children should experience a varied diet of investigations. The following give examples of other different investigation types:

(1) **Classifying and identifying,** eg which materials are magnetic? What is this small creature, its name and type?

(2) **Exploring.** Observations of an event over a period of time, eg germinating seeds or the development of frog spawn.

(3) **Pattern seeking.** Observing, measuring, recording then carrying out a survey to see if there is a pattern, eg do tall people have the longest legs?

(4) **Making things or developing systems.** This maybe a combined science/technology investigation, eg designing and making a traffic light system.

(5) **Investigating models,** eg does the mass of a candle increase or decrease when it burns?

There are many ICT opportunities that can be used in investigations eg data handling.

Many non-investigative activities may be turned into investigative ones if it is appropriate to do so by changing the task title. For example an illustrative activity such as making carbon dioxide using Alka-Seltzer and water can be made into an investigation by posing the question, 'does the amount of Alka-Seltzer used affect the amount of carbon dioxide made?'

USING ICT TO ENHANCE TEACHING AND LEARNING

Today's scientists model, measure and report on their experiments and they do this using a variety of ICT software. Children developing their scientific understanding also need to do these things. As they work they need to record their ideas, assemble data, perform measurements and communicate their thoughts and findings to others. In a modern world 'content free' ICT tools, ie word processing/spreadsheet packages etc, are crucial to this process. The use of such software allows the young scientist to focus on the science involved and not on elaborate drawing or time-consuming colouring-in of graphs.

To this end, this book has pointers, in the margins, towards opportunities for the use of ICT. The icons next to the activities are indications of the opportunities to use generic software to enhance teaching and learning in the contexts indicated. There are many different types of 'content free' pieces of software available that allow users to communicate and handle information to support work in primary science. It is important to be able to distinguish between the types so that decisions can be made about their suitability for specific purposes.

COMMUNICATION

The software tools used for communicating fit into the categories below. All have different functions but most will allow for the communication of information in text, graphics and sound.

- Word processors
- Graphics packages
- Presentation tools
- Multi-media authoring packages

Word processors can store both text and images and allow these formats to be reviewed. They are simple to use, can be accessed quickly and the data in them amended. Modern word processors are also good at handling text, graphics and sound making them useful multi-media tools. Tables built in a word processor are exceptionally useful for clarity of collation of information. It is possible to use a conventional word processor with a 'find' option to do some careful 'key word' sorting of collected data.

Graphics packages allow the creation and storage of images. Each child drawing a picture from observation and then saving it so that the images can be reviewed and re-ordered would be a good KS1 model.

Presentation packages such as MS PowerPoint allow for the input of text and images and these can be easily ordered and displayed. Sound and video can also be added here. It is important to remember that data is not just text and numbers but can be sound, still image, moving image and animation.

Multi-media authoring packages allow for many different kinds of information to be combined together to make a sensory experience out of searching and accessing information. The World Wide Web and the CD-ROM are manifestations of this idea.

DATA HANDLING

The software tools used for data handling fit into the following categories:

- A 'free text' or 'card file' database
- A branching database
- A graph maker
- A flat file or single file database
- A spreadsheet

Each does a specific job but there are overlaps and some versions of each are more suited to younger than older children.

A 'Free text' database

These databases allow users to type in information about things or collect sets of pictures of things in their own way. The database itself is unstructured so can be used by very young children. There are proprietary versions of this type of software, often called 'card files'.

A Branching database

Branching databases (or sometimes called binary trees) are those which work by the users supplying 'yes' or 'no' answers to a series of questions. There are two main types in use at present but the principle of each is based on the idea of unique identification. The usual way of building such a database is to select a number of items and proceed to write questions which divide the items up based on the 'yes' or 'no' answers to carefully written questions. The most general questions are asked first and the more specific ones later. The process of doing this eventually identifies each individual item in the database. The valuable thing, in terms of science in the primary school, about using this type of database, is in getting the children to build the trees themselves. The technology is a way then of displaying their thinking and allows them to refine it.

A Graph maker

This is a software programme which simply allows data collected to be presented in a graphical format. Some programs present the data as pictograms, as ticks (as in a tally chart), as blocks (as in the manual version – unifix or multi-link) or as standard bar, line and pie charts.

A Flat file or single file database

This type of database stores information in records which have a common format, being divided into a number of pre-determined fields. Before data can be entered into the database it must be 'set up' to receive it. The decision to use a flat file data base usually begins because there is a question to answer that needs the collection of some data eg do all the materials we have tested stretch? The information can then be viewed graphically. Flat file databases are good with words and numbers but are better for words.

A Spreadsheet

A spreadsheet is like a flat file database in many ways. Data is stored under similar field headings but can be manipulated in many different ways. Spreadsheets are excellent at handling numerical data, they are also good for mathematical modelling and doing calculations. Information in them is arranged in cells, and any information – numbers, words, formulae etc. can be entered into an active cell. The cells can be selected and the information in them displayed graphically in any of the usual ways (block, bar, pie, line etc.) including correlations with scattergrams.

USING ICT

SENSING PHYSICAL DATA (DATALOGGING)

With a sensor connected to a computer via an interface box or connected directly through one of the available ports (this depends on the sensors), physical data such as changes of temperature, light or sound can be collected and displayed. The simplest sensors display the data in spreadsheets or in software written to optimise visually the information, which has been collected. In science at a primary level probably the most useful of these is a sensor which can record temperature change over a period of time or as 'snapshots'.

There are sensors available that can be used independently of a computer, and then connected for data analysis.

CONTENT SPECIFIC SOFTWARE TO SUPPORT WORK ON MATERIALS AND THEIR PROPERTIES

There are numerous CD-ROMs and programs that could be used by teachers to capture the interest and imagination of their pupils. The use of these powerful tools as motivators and catalysts for discussion is the responsibility of individual teachers. It is they who must do a little research to ensure that the materials presented to children do not harbour the potential to bring about misconceptions, nor does the excitement of ICT detract from the learning of science.

Reviews of software available for primary science education can be found at http://www.chemsoc.org/networks/LearnNet/thats-chemistry.htm

Below is a very short list of content specific software which teachers might find useful for enhancing the work on materials and their properties.

Science Explorer	Granada
Science Explorer II	Granada
Becoming a Science Explorer	Dorling Kindersley
Mad About Science 3 Matter	Dorling Kindersley

USEFUL WEBSITES

This is not intended to be an all-inclusive list. Some of these sites are for children with information at that level and some are for teachers to improve their subject knowledge and to support ideas. Most will have some content for both purposes. Teachers will need to visit these sites and spend time deciding whether the information/simulation/activity displayed enhances their teaching and the children's learning. (Sites accessed November 2000)

chem4kids
http://www.chem4kids.com/

e4S have some animations
http://www.e4s.org.uk/

QUEST
http://www.nhm.ac.uk/education/quest2/english/index.html

Scienceweb
http://www.scienceweb.org.uk/index.htm

Virtual Teachers Centre
http://www.vtc.ngfl.gov.uk/vtc/curriculum

Volcano world
http://volcano.und.nodak.edu/vw.html

Science Museum
http://www.nmsi.ac.uk

Roger Frost's site
http://www.rogerfrost.com

TOOLKITS

Many software companies put together generic toolkits. Sets of software which integrate together to provide users with the ability to communicate in text, graphics, charts and tables.

It is worth contacting RM, Granada Learning, BlackCat Educational Software, Softease and Edu Tech, amongst others, for details of these integrated packages. Contact details for these companies can be found at

http://www.chemsoc.org/networks/LearnNet/thats-chemistry.htm

this page has been intentionally left blank for your notes

GROUPING AND CLASSIFYING MATERIALS
on the basis of their simple properties

Science background for teachers

VOCABULARY

Names of a variety of materials; wood, metal, plastic, paper, rock, sand, chalk, fabric, leather, cotton, oil, wax, natural, man-made
Words to describe materials using the senses; shiny, dull, transparent, soft, hard, rough, smooth, including 'smelling' and listening words such as smells 'peppery', sounds loud
Words to describe properties; solid, liquid, strong, tough, magnetic, bendy, squashy, elastic, waterproof, floater, sinker

For children, 'materials' often mean fabrics or textiles when in fact we can use it to mean any substance. Materials exist as solids, liquids or gases and as mixtures of these and when introducing their classification it is worth beginning with this grouping especially with young children, since, for example, many of them think of liquids as being only water. Materials are also natural or man-made. Confusion often arises for children because man-made materials can be divided into two groups; those that are derived from natural products but are refined or altered by man for his use, and synthetic products originally derived from substances from the Earth and then changed chemically into new products. A further complication comes from the fact that some things originally made from the natural material like candles from beeswax, are now more commonly man-made, in this case from paraffin wax. At this stage it is also worthwhile discussing the term man-made where the word man is used generically and not specifically.

Natural materials include wool, cotton, linen, leather, wood, cork, stone, gravel, sand, salt, coal, gypsum, talc, some metals eg gold and silver, silk, oil, gemstones, beeswax.

Converted raw materials include pottery, china, earthenware, most metals eg steel and aluminium, coke, charcoal, rubber, paints, some medicines and drugs, paper and viscose.

Synthetics include plastics, polyester, acrylic, PVC, nylon, polythene, glass, some other medicines and drugs.

Materials are used for different jobs on the basis of their properties, so young children need to begin to identify such properties before they can consider the suitability of materials for different uses. They can group materials on the basis of the simple properties that they can experience with their senses, beginning with simple ideas such as texture, and building up to more complex concepts, such as elasticity. A large amount of descriptive vocabulary can be introduced in this context.

Most towns and villages have areas where glass, paper and metals can be taken, sorted and put into the appropriate bins for recycling. This science topic is an ideal opportunity to raise the children's awareness of this idea.

SKILLS
- Describing with increasing accuracy using the senses.
- Grouping according to different and given criteria.
- Carrying out simple tests.
- Working cooperatively.
- Recording in different ways.
- Use software to combine words and pictures about objects.

GROUPING AND CLASSIFYING MATERIALS

Key ideas and activities

The senses can be used to explore the differences and similarities between materials

(a) **Tactile** Prepare a 'feely' bag with different objects inside which have a variety of tactile properties. Pass the bag round, let the children feel an object without looking and describe what they feel, then pass the bag on. Include bendy, stretchy and squashy objects.

(b) **Observation** Children draw some objects and describe them; include transparent and shiny objects. Encourage the children to name the material the objects are made from.

(c) **Sense of smell** Prepare some 'smelly' materials and put them in opaque jars or boxes with lids. The children take off the lids and describe the smells without looking inside. (Use small pieces of soap, spices, onion, herbs, cotton wool soaked in vinegar, lemon juice, coffee.)

(d) **Hearing** Show the children a tray of objects, then unseen you tap the object (set up a screen or get the children to turn round) and the children guess which one it was and describe the sound. Include soft items such as a cushion so that they hear nothing even though you are hitting it. (Noise pollution is often a consideration in using a material eg rubber feet on furniture.)

(e) **All the senses** Give the children different objects which they describe using all their senses.

(f) **Sorting materials using the senses** Children can begin to classify and sort objects and the materials from which they are made. Set up different tables displaying different types of materials eg a table of shiny things, transparent things, soft things, and get the children to record the object, material and its property.

Include a table of materials that may be stretchy or squashy, older children can be introduced into the idea of materials being 'elastic'.

Communications in text/graphics

Object	Material	Properties
Bear	Wool	Soft, Warm, Squidgy
Car	Metal	Hard, Cold, Shiny

Toys on our table.

GROUPING AND CLASSIFYING MATERIALS

(g) **Materials may be sorted in various ways** Re-sort some of the materials in a different way eg a glass is transparent and hard.

(h) **Recognising materials/objects using a verbal description** Play an 'I spy' game, describe an object or material and the children guess what it is from your description.

ART ACTIVITIES such as drawing, painting and using materials to make collage can be used here very effectively. For example, get the children to make collage using only paper, only shiny materials, only natural materials etc.

Recognising, naming and sorting common types of material including those which are found naturally

(a) **Naming and sorting** As in activity (f) above set up tables of different, named materials – wood, metal and plastic, and get the children to record in this way. They can begin to do Venn activities using hoops to sort the materials and eventually recording using Venn diagrams. At this point it is worth discussing how objects do not always look as though they are made from the same material, even though they are, and this is because of the variation within the same material, eg different types of woods and plastics. Older children can be introduced to specific names of materials, such as polythene bag instead of plastic bag, stainless steel rather than metal.

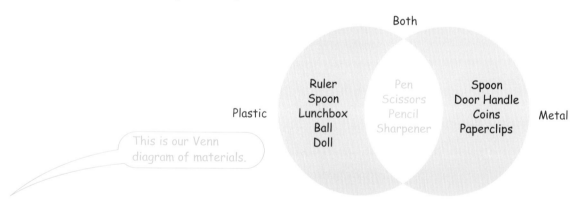

This is our Venn diagram of materials.

(b) **Matching card game** Make or purchase a matching materials card game for the children to play, where children match a picture with the name of a material that it may be made from. So they may match a picture of a bucket with 'plastic' or 'metal'.

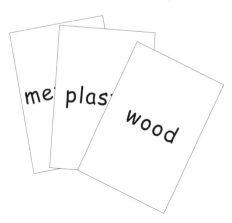

We made a matching card game about materials.

GROUPING AND CLASSIFYING MATERIALS

(c) **Natural and manufactured materials** Begin to discuss the idea of natural materials using obvious examples such as wood as natural and plastic as man-made and get the children to record some and say their origin, eg draw a wooden object and a tree. A sorting activity could be done also using natural and man-made as the criteria. Older children can consider sorting man-made products into synthetics (raw materials that have been chemically changed such as nylon) and non-synthetics (converted raw materials such as pottery).

Materials have a variety of uses and the same object may be made from different materials

(a) **Different uses for the same material** Using for example, a 'wood' table, discuss the uses of the different objects made of wood and get the children to record these or their own objects or collect pictures and record the uses. Do this for a variety of materials.

(b) **Objects used for the same purpose made from different materials** Look at a collection of similar objects eg mugs and discuss what they are made of. Do this for some other objects, eg shoes, hats.

Objects and materials have different properties that can be tested, and may be grouped according to those properties

Some of these activities may be set up as simple classification investigations by asking the children how they would find out which things were for example magnetic or floated. Older children could investigate Are all metals magnetic? Which is the best magnet?

(a) **Ability to float** Get the children to test a range of objects and materials using bowls or plastic boxes of water. Get them to change the shapes of some materials to see if it makes a difference, eg foil and plasticine screwed up and made flat.

Branching database
Venn diagrams

(b) **Magnetic or non-magnetic** Give the children magnets and a selection of objects including metals and non-metals to find out which are magnetic. This can be used as a simple classification activity which they can record in chart form.

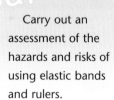

Safety!
Carry out an assessment of the hazards and risks of using elastic bands and rulers.
Goggles should be worn.

(c) **Ability to change its shape** Which materials will change their shape and remain in the new shape (plastic) or return to their original shape (elastic)? Give a variety of materials to test including, clay, (part of an art activity), plasticene, Blu-tack®, foam sponge, elastic bands, nylon, rubber ball, plastic rulers.

Object	Material	Change Shape	Go back again
Ball	Rubber	Yes	Yes
Lump	Plasticene	Yes	No

d) **Waterproof** Which materials are waterproof? A simple way to carry out this activity is to lay different materials onto paper and, using a dropper, drop water onto the materials. If the water goes through, the material is not waterproof, if it sits on the top it is. Alternatively, stretch equal sized pieces of material over jam jars and drop equal quantities of water from a dropper onto the fabric. Water will sit on the top of waterproof materials. A chance to discuss 'fair testing'.

e) **Absorbency** If a material is not waterproof then it may be absorbent and sometimes absorbency is a quality that we want, for example, in blotting paper and babies' nappies. Absorbent materials will hold a lot of water in their fibres. The waterproof activity above can also be used to test for absorbency. Instead of dropping equal quantities of water onto the fabric, water is dropped on continually until the fabric will hold no more and the water just drips into the jar underneath. The amount of water absorbed by each fabric can then be recorded. The most absorbent fabric is the one that holds the most water. Older or more able children could plan their own investigation; 'Which is the most absorbent cloth?'

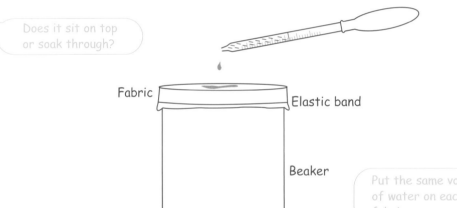

Does it sit on top or soak through?

Fabric — Elastic band — Beaker

Put the same volume of water on each fabric.

Safety!
- Goggles should be worn when testing some elastic materials eg elastic bands, and plastic rulers.
- Care should be taken when handling glass.
- Collections of objects may present potential hazards – sharp, small and may be swallowed, etc.

f) **Making waterproof paper** This activity can have cross-curricular links with art. Rubbing or drawing with wax crayon or candle onto paper then washing over the picture with a paint or ink wash. Children can see that the wash does not go where the wax has been.

g) **Drying fabric** Investigate Which fabric dries the fastest? An opportunity to involve fair testing using equal sized pieces of fabric, measured equal volumes of water, timing, and recording results in a block graph. This is also an opportunity to use science in an example of an everyday activity.

GROUPING AND CLASSIFYING MATERIALS — RS•C

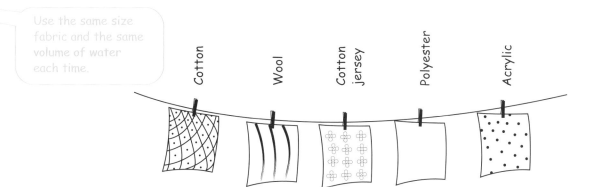

Use the same size fabric and the same volume of water each time.

Where shall we put our washing line?

A graph to show the time taken for fabrics to dry

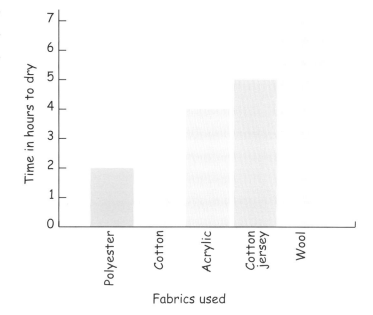

Graphing package

h) **Transparent papers** Which papers can you see through clearly (transparent), which show the light through (translucent) and which show nothing through (opaque)? Children can investigate papers and put them in an order of transparency. Older children can learn the correct terms.

i) **Writing/drawing paper** Is all paper good for drawing on? Investigate which is the best. This activity is beginning to look at the specific uses of a material for a purpose, and is very appropriate for young children. All children have experienced the

Word processing

Writing Tool	Type of Paper			
	Cartridge	Tissue	Tracing	Newsprint
Lead Pencil	Good	Tears it	Good	Faint
Coloured Pencil				
Wax Crayon				
Pastel				
Chalk (White)				

frustration of paper that you can't draw on properly. This investigation is simple, keeps the variables to a minimum and they can use the paper they test as a record for their results. They could also try different writing implements on different paper. The children can chart their results.

Woolly Saucepan The wrong materials used for things, as the name suggests.

Bottle Bank At the bottle bank, things are sorted according to the colour they are.

Grandparents House In the house there are lots of interesting things made from different materials.

Scoop a Gloop Clay is a very tactile material, which also has a distinctive smell. It can change its shape and finally becomes hard.

Night-time Kitchen The different materials used in the kitchen.

Rubber Dubber Balls are made of various materials, this one is rubber and bounces.

star* Poetry

by Michael Rosen

Woolly Saucepan

Could I have
a woolly saucepan
a metal jumper
a glass chair
and a wooden window-pane please?

Er-sorry – I mean
a woolly chair
a glass jumper
a wooden saucepan
and a metal window-pane please?

Er-sorry – I mean
Oh – blow it!
You know what I mean, don't you?

Bottle Bank

The bottle bank
gobbled up my bottle
and the bottle bank
went clank.

That's bad. Look at all the work I do
giving it hundreds of bottles to chew.

And that's not all:
at home,
we've got three bins:
one for bottles,
one for paper
and one for tins.

After all that work
I don't think
a bottle bank
should just say clank.
I think
bottle banks
should say thanks.

Grandparents' House

At my grandparents' house
they've got a very old plate
with a gold edge round it
and my grandad says,
it's real gold
that's so thin it's like paper.

At my grandparents' house
they've got a very old picture
made of wood,
and my grandma says
that her grandad made it
by cutting out lots of tiny bits
of different kinds of wood
to make all the colours.

At my grandparent's house
they've got a very old newspaper that tells the story of how
when my grandparents
were teenagers
they once rowed out to sea
and nearly drowned
and the paper has gone all brown.

At my grandparents' house
they've just fitted a new kitchen
and it looks like it's made of wood
with all the lines and whirly bits
but really it's plastic.

I love going to my grandparents' house

and looking at all their stuff.

Scoop a Gloop

Scoop a gloop
of slimy clay
squeeze it, knead it,
pummel it, stretch it
roly poly, roll it
into long, thin
sausages.

Bend them, coil them
one on top
of one another
up and up
and round and round
to make a
pot.

It's still soft
and leans a bit
but wait –
and wait –
it slowly hardens
sits dry and dusty
crisp as a biscuit.

Don't tap
or drop
it'll crack
or crumble.

Take it gently
to the kiln
and under fire
of fantastic heat
it strengthens
toughens
enough
to let you
use your spoon
or run your
thumbnail
up and down
your clever coils.

GROUPING AND CLASSIFYING MATERIALS

Night-time Kitchen

It was all dark in the kitchen
Everyone was in bed,
when suddenly the saucepan said
'It's time I had a bit of respect
around here.
I get thrown about, banged down,
scraped with a spoon,
left for hours covered in old food.
I am made of the finest steel.
I want everyone to know
that if it wasn't for steel
and all the other metals round
here this whole place would
grind to a standstill.

Without us, there would be **nothing**.
We are the most important.
They wouldn't be able to cook
without their metal cooker.
They wouldn't be able to eat
without their metal knives and forks.
They wouldn't be able to drink
and keep clean
if it wasn't for all the metal pipes.

From now on
everyone round here
should call us Lord.
Lord Saucepan, Lord Spoon,
Lord Tap and – '

The breadboard had been listening
to all this
and was getting cross.
 'Hang on there, Potty!
Those of us round here
who are made of wood
think we've got a case.'
'Huh!' said the saucepan,
'hark at old Blockhead!'

'No, listen.
Without wooden table and chairs
they'd be eating off the floor.
But without wooden floorboards
they'd be eating off the ground.
But without the wooden beams
the house would fall down
on everyone
so no one would be left alive
to use you, Mr Potty.
If anyone round here ought
to be called Lord
it's people like Floorboard.'

'So,' said the saucepan,
turning to the window,
'who's the most important
   round here?
Metal or Wood?'

And the window said,
'This is crazy.
We don't think either of you
should boss over the rest of us.
You're both great stuff –
different but both great.

But watch it –
You saucepan. They're making
glass saucepans, these days.
And you, table!
Glass tables are really rather fancy.'

And at that
metal and wood
agreed to respect each other
though they're still arguing
they would stop rowing
though they're arguing over
which of them
should be the door handle!

Rubber Dubber

Rubber dubber
flouncer bouncer
up the wall
and in and outer
under over
bouncing backer
mustn't dropper
mustn't stopper
in betweener
do a clapper
in betweener
do a spinner
faster faster
to and fro-er
rubber dubber
flouncer bouncer

BUT

then oh bother!
Butter finger
dropped the ball
and pitter patter
patter pitter
rubber ball
ran right away.

THE PROPERTIES OF MATERIALS
and their everyday uses

Science background for teachers

VOCABULARY

Words to describe properties: Hard, soft, strong, weak, tough, brittle, stiff, rigid-flexible, absorbent, waterproof, magnetic, non-magnetic, wear and tear, smooth, rough, transparent, opaque, translucent

Names of a variety of materials: wood, metals – copper, tin, steel, gold, silver, aluminium, chrome, plastic – polythene, polystyrene, PVC, fabrics – cotton, silk, polyester, wool, acrylic, foam, glass, rubber

Children need to have experience of, and explore as many different materials (substances) as possible in order to make sense of their world. Understanding how materials behave in their natural state and under certain conditions will help them to understand why objects are made of specific materials. Some properties are easily observable features, such as transparency, which they explore as younger children, others are less obvious and need to have tests carried out on them.

In carrying out comparative tests on different materials, children develop an understanding of suitability for different purposes. They then begin to develop the skills themselves to choose the best materials for certain tasks. When testing materials for properties, precise vocabulary becomes important because children (and adults) sometimes confuse scientific terms, which they use in a general way in their everyday speech.

Hardness   Resistance to scratching and pressure. Hardwood does not mark as easily as softwood.

Strength   Amount of force needed to break a material usually by pushing or pulling down.

Toughness   Resistance to breaking by cracking, opposite to 'brittle'.

Stiffness   Amount of force needed to change the shape of a material, opposite to flexible.

Elasticity   Ability to return its original shape when a force is removed eg rubber band.

Plasticity   Ability to retain the new shape when a force is removed eg plasticene.

Absorbency   Ability of a material to soak up a liquid.

Waterproof   Resistance to liquids, repels water.

A material can be described in a variety of ways for example it may be strong but brittle, and the combination of its properties may determine its use. The property of a material can change according to how the material is treated; clay is very different once it has been fired, rolled up newspaper is very different to a sheet of newspaper.

THE PROPERTIES OF MATERIALS

SKILLS
- Recognising and carrying out a fair test, repeating a procedure.
- Measuring length with a degree of accuracy.
- Constructing a bar graph.
- Careful observation.
- Working cooperatively.
- Recording carefully.
- Use of ICT for graph drawing.

THE PROPERTIES OF MATERIALS
Key ideas and activities

Younger children should have spent time experiencing some testing of the simple properties of materials. The activities offered here try to build on that experience and give the opportunity to develop the skills of investigating, whilst tackling the testing of properties. Certain materials have properties that are appropriate for specific uses and by comparing these, the idea is introduced that whilst some materials are reasonable for the use, others might do the job better.

Some materials can be changed in shape, a property suited to specific uses

Graphing package

(a) **Balls are made from a variety of materials** Investigate
Which is the bounciest ball? Look at balls made from a variety of materials and discuss the different uses and properties, including sizes. The children can decide what they want to test eg, 'the bounciest football', 'the bounciest small ball'. What do they think 'bounciest' means? A ball that bounces the highest, or one that bounces for the longest time? The possibilities and variables are numerous, so they need to be made more specific. This is where the children learn to plan. The results will make a good bar graph.

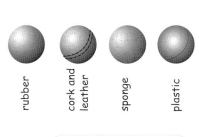

Choose balls that are the same size. How high did they bounce?

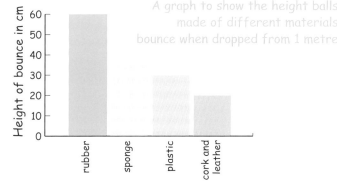

A graph to show the height balls made of different materials bounce when dropped from 1 metre

- Goggles should be worn when testing elastic bands.
- Care should be taken when using 'scratching' tools.
- Care should be taken when using weights.
- Great care should be taken and children closely supervised when using hot wax.

(b) **Elastic bands** Investigate elastic bands with older children. Stretching them to their breaking point is too dangerous! Try comparing different thicknesses of the same band circumference with a fixed weight eg 500 gram. Carry out this experiment on the floor. Hold the band and weight against a ruler to see which stretches the most.

Testing different thicknesses of rubber bands with weights

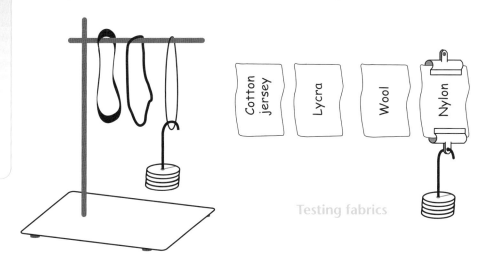

Testing fabrics

THE PROPERTIES OF MATERIALS

(c) **Stretchy materials** Sometimes we want a material (fabric) to have some elasticity because of the garment it will be used to make, for example, a pair of tights. Different fabrics can be tested for this, but pupils must remember that we want the fabric to return to its original shape and size. Strips of fabric can have weights hung onto them. What length is the fabric at the start? To what length does it stretch? What length does it return to? Use bulldog clips to support the weights or cut a hole in the fabric to hang the weights through.

Spreadsheets

(d) **Flexibility** Some usually rigid materials need to be able to 'give' a little and not break, to accommodate different situations for example a bridge carrying heavy traffic. Different materials can be tested eg identical lengths of wood, plastic, metal (use rulers) and card to investigate how much they will bend by hanging weights from string onto the end or sticking weights on the top with Blu-Tack®. Since the intention is not to break all your material samples, use a light weight and investigate 'which material bends the most using a 100 gram weight?' Measure the distance that each ruler bends. Another consideration is the way the material is formed. A card tube, for example, is less flexible than the same card unrolled. Children can investigate one material in different forms. Cut the card from both sides of a cereal packet to test the card flat, rolled the long way, the short way and folded zig-zag, then put the weights on the top.

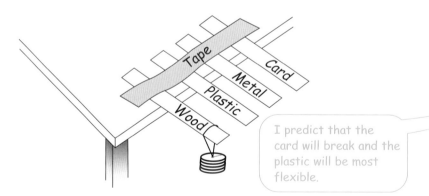

I predict that the card will break and the plastic will be most flexible.

Comparing everyday materials for strength

Most materials need to be strong. They need to withstand the forces of pulling or pushing without breaking or tearing and withstand 'wear and tear' if they are materials to be made into clothing.

(a) **The strongest paper** Papers are made of different qualities, young children may have carried out the simple test for the best drawing paper. This is another one for younger children. Paper is often used for wrapping parcels and needs to be flexible and strong. Investigate Which type of paper is strong? Can they think of a way to test the strength of paper? Collect sheets of different types of paper and make them the same size. Make a hole in each sheet not too near the edge and hang a weight carrier onto it. Carefully add weights until it tears. A sheet of A4 computer paper held 650 g before tearing.

Tip If weight carriers are not available use a strong bulldog clip with a pot attached to it by thread going through the hole. Clip this to the sheet of paper and gradually add weights to the pot. The bulldog clip sometimes slips off the paper, so wrap the end of the paper around a pencil and the bulldog clip will grip this.

Wrap the end of the paper around a pencil to stop the bulldog clip from slipping off.

Make a bridge with paper Older children can try making a bridge using different types of paper and putting weights on the top. What happens if the paper is folded into a concertina shape?

(b) Testing threads Many fabrics are made from woven or twisted threads and their strength is important. Strings and ropes made from such fibres may hold heavy weights, peoples' lives may depend on their strength. Investigate the strength of threads by suspending weights from equal lengths of threads and seeing the weight they will hold until they break.

Tip Weight carriers can be used, but threads are surprisingly strong, a length of cotton held 1100 g, polyester and nylon will hold much more. Try tying the threads to the leg of a table turned on its side, tie a small, light plastic bucket (plasticene tub) to the thread being tested and gradually add weights. The container allows large weights to be added. Threads may also be tested using a newtonmeter, (forcemeter), but it is less accurate as it is often difficult to see exactly where the thread breaks because the spring returns too quickly.

(c) Testing fabrics for strength Fabrics used for items such as work clothes, sheeting, bags etc need to be strong to withstand being pulled around, for example by your friends! The strength of fabrics can be tested in various ways, similar to the testing of threads. See if the children can suggest ways in which this might be done. Equal strips of fabric can have a newtonmeter attached to the end. As it is pulled, the force will register on the scale to the point where the fabric tears. As fabric tears slower than thread breaks, this method works quite well.

Tip Alternatively, by using a pot as the container or a weight carrier, the weights can be hung through a hole in the fabric until it tears (some fabrics are very strong and can take many weights, use thin fabric strips). Investigate 'Which is the strongest fabric?' You may discuss with the children whether it is fair to include 'elastic' fabrics in this test that also have the 'stretch' property.

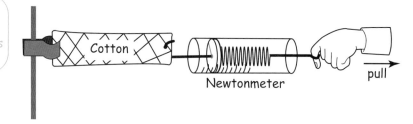

Slowly pull the newtonmeter to see how much force it takes to tear the fabric.

(d) **Hard wearing fabric** Clothes must withstand the wear and tear that comes from sitting on the floor or playground tumbles, (who has had tears in their trouser knees?) rather like friction wear. Discuss this with the children and see if they can think of a way to test this. If you stick a piece of coarse grade sandpaper onto a wood block and rub this onto fabric you can count how many times you can rub before the fabric shows wearing or a hole. Choosing the same child to rub makes it fairer, although a little tiring!

(e) **Which is the best carrier bag?** **Investigation** for older children. Children bring in a collection of carrier bags. What do we mean by the best? The strongest? Handles that don't mark your fingers or both? Bags may be tested, by hanging them over a pole (or the leg of a table turned on its side). Weights are then carefully put into the bags (they take a lot, supplement with bricks). The 'best' bag will carry the heaviest load. A separate **investigation** may be carried out to test for handles that don't mark your fingers but older children can deal with both variables at once. Whilst testing for strength, put plasticene on the pole where the handles are to be hung. The handles cut into the plasticene as the bag becomes heavier. The amount the handles cut into the plasticene can be measured. The results can then be discussed. Which is the bag that is the strongest but marks your hands the least? Is there a compromise to be made?

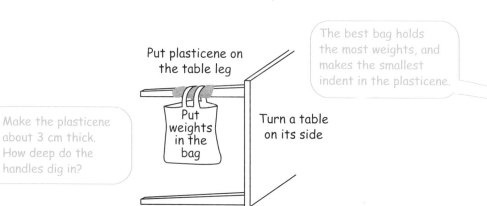

The best bag holds the most weights, and makes the smallest indent in the plasticene.

Make the plasticene about 3 cm thick. How deep do the handles dig in?

Comparing everyday materials for hardness

(a) **The scratch test** This test, which is used on different rocks (see Rocks and Soils section) can also be used on other materials such as different woods or different types of hard flooring such as vinyl, linoleum and tiles. Discuss with the children the fact that flooring must withstand scuffs and scratches from shoes and scraping chairs and see if they can devise a test. Scratching can be done with a variety of objects from fingernails to metal nails on samples of the material to be tested. Discuss with the children the difficulty of measuring how hard to scratch. If the same child scratches each material as hard as possible it helps to make the test fair. Children will often come up with this conclusion themselves.

Investigate Which is the hardest wood?

(b) Denting test If you drop a metal weight on to a piece of flat plasticene it will leave a dent. What will it do to the floor? There are often notices in halls where there is a wooden floor about wearing stiletto heels. There are a variety of variables here for older children to consider apart from the material that is to be tested, the different weights, the different heights they may be dropped from and the different surface area of the weight. Drop weights through a wide drainpipe or cardboard tube for safety, onto squares of different types of flooring, cork, vinyl, carpet, wood and ceramic. Furniture often leaves marks on carpets and flooring, so you could test four pieces of flooring. Place different flooring under the four legs of a classroom chair and weight the seat of the chair with books or bricks. Leave for a few days then look to see if there are any marks.

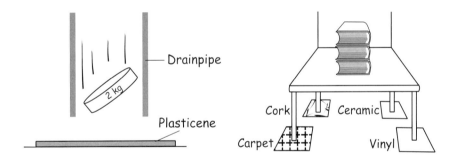

Tip Ask the children to bring in samples of flooring from home since parents often keep 'off cuts' or try to get old samples to keep from a local builders merchant or DIY store.

Comparing everyday materials for absorbency

Kitchen paper and disposable cloths are common items in most kitchens these days for doing many jobs, especially mopping up spills, but are all makes of paper as good as each other? Are some materials better than others? When carrying out these tests the children must consider how much water to use, the size of the material, how many pieces/wipes they use. A pipette or syringe dispenses accurate, small quantities of liquid, which can be coloured with vegetable dye for some of these tests.

(a) Mopping up Young children might drop a small, measured quantity of water onto a tray and put sheets of paper onto the top until it is all absorbed. Alternatively, drop measured quantities onto the paper. Investigate which is the best kitchen paper?

(b) **Which material absorbs water the fastest?** Lay thin strips of equal lengths of different materials (include a waterproof strip) into a shallow tray. Pour coloured water into the tray, measure how fast and how far up the fabric it is absorbed. Discuss the different reasons why absorbent materials may need to be used.

Put thin strips of fabric in a dish of coloured water

(c) **Absorbent building materials** It does not always occur to children that 'hard' materials like wood and stone can absorb water. The knowledge of this fact is particularly important in understanding how building materials (or their toys!) need protection from the wet. Children can test a variety of materials eg plastics, metals, different types of wood and bricks for their absorbent property. Young children can put small objects eg tennis ball, wooden brick etc into very shallow dishes of coloured water to observe if any water is absorbed. Older children can use a dropper to measure out 5 cm^3 into each dish. Stand small pieces of material of the same size in the dishes eg cork, wood, metal and plastic to observe and time any absorption that might take place. If accurate scales are available, a quantitative and 'fair test activity' could be to investigate which is the most absorbent brick or wood. Compare different types of bricks or wood of the same size, weigh before and after immersing in water (see also Rocks and Soils section). A bar graph can be constructed showing the volume of water absorbed by each type of wood.

(d) **Waterproofing** If materials do not absorb water at all they are said to be waterproof, a property that can be very useful for example in clothing, packaging and housing. A very simple activity for younger children is to drop water onto a variety of materials as in the 'absorbency' activity, to see if they absorb or repel the water. Older children can stretch pieces of cloth across jars to make them taut and using a dropper, drop water onto the cloth. The number of drops and the time they stay on the cloth before being absorbed, if at all, can be measured. Making the absorbent material 'waterproof' can then be tried with older children, by waxing it.

THE PROPERTIES OF MATERIALS

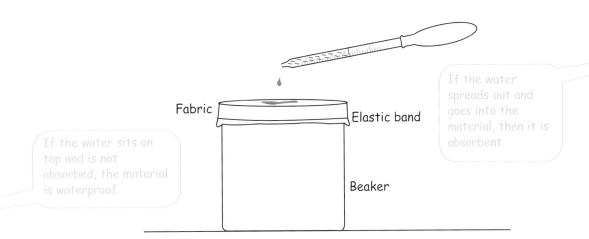

If the water sits on top and is not absorbed, the material is waterproof.

If the water spreads out and goes into the material, then it is absorbent.

Fabric — Elastic band — Beaker

Batik The ancient craft of dying material using a wax resist could be done as an art activity. Molten wax is applied onto cotton through a special tool called a tjanting or painted with a brush. The cloth is then dyed and when it is dry, the wax is ironed out. Where the wax has been applied, the dye does not penetrate, so a pattern is formed. This is an extension of the wax drawing and wash technique often used with younger children.

(e) **Make a waterproof hat/bag** A design technology activity.

Comparing everyday materials for magnetism

(a) **Is it magnetic?** Younger children may carry out a simple classification activity to see which materials are magnetic.

(b) **Will magnets work through all materials?** This can be a simple classification activity or investigate which materials magnets work through the best? Does thickness matter?

Branching database

(c) **Which is the best magnet** Using different types of magnets get the children to plan and investigate this.

(d) **Make a magnetic game** A design technology activity to make a magnetic game, with instructions and rules.

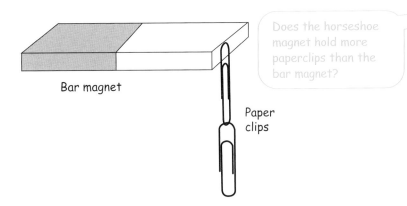

Does the horseshoe magnet hold more paperclips than the bar magnet?

Bar magnet — Paper clips

MATERIAL FACT FILE To 'round off' and consolidate this work, or as an introduction or research for older children (homework) here are some suggested 'pencil and paper' activities:

THE PROPERTIES OF MATERIALS

(a) Choose a large object with many materials and functions eg a bicycle, room in a house, outside of the house and get the children to construct a chart of the materials used, the function and properties of the material. For less able children provide a copy of the sheet at the end of the chapter.

(b) Choose a material with many different uses and properties such as aluminium, write them down or let the children research for themselves and match a particular property to a specific use eg flexible, lightweight – aluminium foil. Non-corrosive, lightweight-bicycle frame. For less able children provide a copy of the sheet at the end of the chapter.

by Michael Rosen

Bag Words A poem that plays with words, what are bags made of?

Scary Sausage-fingers That's what carrier bags do to your fingers if you have lots of shopping.

Night-time Kitchen Which is the most important material in the kitchen? See page 10.

Rubber Dubber Rubber is 'elastic' and used for making bouncy balls, what else is it used for? See page 10.

Bag Words

Today we are learning about bags.

If a paper bag is made of paper
is a handbag made of hands?
is a sandbag made of sand?

An air bag is full of air
so I suppose a plastic bag
is full of plastic.

A carrier bag carries,
so a sick bag is sick.

I know what mailbags
look like
but what do bags of fun
look like?
And can you pack the bags
under your eyes?

I know who let the cat
out of the bag
but who put it in?
I just hope there aren't any cats
in the bags under your eyes!

I've heard there's a
bag of nerves
and a
bag of bones.
Why not put them

in together
with some blood, muscles and skin
and you could have
a bag of person?

All this I understand
But why do people
keep saying
'It's in the bag'?

What's in the bag?
The cat?
The sick?

And while I think about it
which bag is 'it' in?
One of those bags of fun, perhaps?

Shouldn't they say
what they mean
and instead of saying
'It's in the bag'
couldn't they say
what they are talking about
like
'The hand is in the handbag'
or
'The cat is in the bag of nerves'.

And then we'd all know
what they're talking about.

Scary Sausage-fingers

Hey!

Psst!

I got a trick.

Do you want to know my trick?

I'll tell it to you.

It's called Scary Sausage-Fingers.

I'll tell you how you do Scary Sausage-Fingers.

You walk into a room
with Scary Sausage-Fingers,
you hold up your hands
and go:
"Look at me! I got Scary Sausage-Fingers!"
and everyone'll go, 'YAAAA!!!'

But first you've got to
make
Scary Sausage-Fingers

You go to the supermarket - without a bag.
You do loads and loads of shopping
You put it all into the bags
they give you
You carry it all home.

As you walk along
the handles of the bags
cut into your hands.

It's agony.

But you don't give up.

You walk on home.

The bags still cutting into your hands.

It's double-agony.

You get home
You drop the bags.

Now you've got
Scary Sausage-Fingers,
Big, fat, puffy fingers
with little narrow white bits
in the joints in between
the sausages.
So you rush into where
everyone's sitting quietly
having a nice time
and you hold up your hands
in the air and shout,
'Look! Scary Sausage-Fingers!'
and everyone'll go, 'YAAAA!!!'

lightweight

expensive to produce

soft, easy to mould and shape

can be mixed with other metals to make it stronger

good thermal conductor

it does not rust

good electrical conductor

Match the Aluminium Facts

Rotary driers

Aluminium foil

Bicycle frames

Saucepans

Electrical cables

Window frames

Toothpaste tubes

Mirrors

Watering cans

Milkbottle tops

Aircraft

Object	Material	Property	Use related to the property of the material
teapot	ceramic	rigid, waterproof resistant to hot liquids	holding hot liquids

Object	Task	Material	Why was it chosen?
sink	holding water, washing things	stainless steel	waterproof, strong, tough, easy to clean

this page has been intentionally left blank for your notes

this page has been intentionally left blank for your notes

INSULATION AND CONDUCTION
thermal and electrical insulation and conduction

Science background for teachers

VOCABULARY

Names of a variety of materials; polystyrene, foam, wool and names of metals-copper, iron, tin, lead and alloys, brass, bronze, silver
Associated words; temperature, thermometer, degrees Celsius, thermal, conductor, insulator, conduct, insulate, predict, measure, volume, cubic centimetres (cm³)

Apart from heating and cooling appliances, objects in a room will be at the temperature of that room, even if they appear to feel warmer or colder. This is a difficult concept for children to comprehend. How hot something feels has as much to do with the thermal conductivity of that object as its actual temperature. Thermal conductivity is the ability to transfer heat from one object to another or away from your hand to the object. Thus wood and metal at the same temperature will feel very different because metal is a good thermal conductor, will conduct heat away from your hand and thus feel cold. Wood is a poor conductor, heat is not conducted away so it feels warm.

Conversely, some materials do not transfer heat easily but slow it down, these are poor thermal conductors but good thermal insulators. These materials keep things warm but also prevent cold things from warming up. As well as wood, air is a good thermal insulator, as is any material that has air trapped in it such as expanded polystyrene, foam, fibreglass and wool. Air molecules are moving around randomly, at a distance from each other so heat energy will not pass from one to another very quickly. Children need to understand that these materials do not actually make us warm but keep the natural body warmth in.

Objects that start out hotter or colder than room temperature because they have been heated or cooled will eventually reach the temperature of the room. For example, hot water and ice-cold water will both reach room temperature if left in a room for an appropriate period of time.

Materials which are part of a working electrical circuit are electrical conductors, they allow electricity to flow. All metals are conductors. Electrical insulators do not allow this to happen. This includes all other common materials, except for graphite (as in pencil leads) which is a conductor.

Good thermal conductors are generally also good electrical conductors,... BUT graphite is a better electrical than thermal conductor.

SKILLS

- Using a thermometer with care and, for older children, understanding minus numbers.
- Accurately measuring volume.
- Recording results in chart form and, for older children, constructing a line graph. ICT can be used here.

INSULATION AND CONDUCTION

Key ideas and activities

Feeling materials is not an accurate measure of their temperature

(a) Give each group of children three bowls of water, hot, cold and warm. The children take it in turn to put one hand in hot and the other in the cold water. After a minute they then put both hands in the warm water. The water will feel different to each hand. They can also take the temperature of each bowl of water as an exercise in reinforcing or teaching the use of a thermometer. Put a piece of metal and a piece of wood in the refrigerator, get the children to then feel them. They feel they are at different temperatures, but are they? A thermometer taped to the material shows that they are the same temperature.

Hot things cool down and cool things warm up to eventually reach the temperature of their surroundings

Have two containers containing identical volumes of water, one very cold and the other one hot. Put thermometers in each and take the temperature at regular intervals until both reach room temperature. This will take some time, and could be done alongside another activity but it is an excellent activity for older children to do and then to construct a line graph. It is also ideal for datalogging if you have the equipment. Compare the final temperatures of the liquids with the temperature of the room.

Datalogging

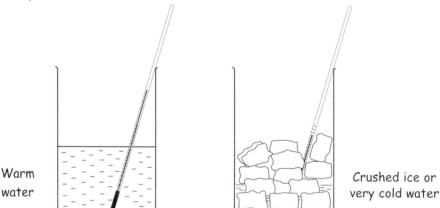
Warm water | Crushed ice or very cold water

Room temperature was 21°C.

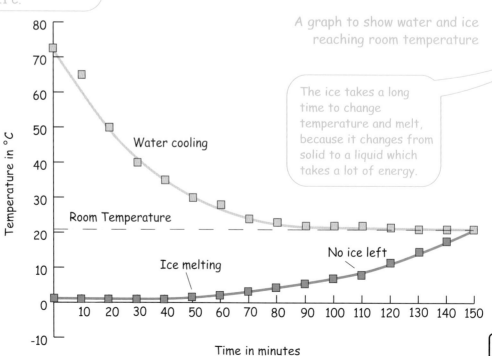

A graph to show water and ice reaching room temperature

The ice takes a long time to change temperature and melt, because it changes from solid to a liquid which takes a lot of energy.

INSULATION AND CONDUCTION

Some materials are better thermal insulators than others

Safety!

Mercury thermometers are not advised for 5-11 year old children, use thermometers that read to 110 °C.
- Ice should not be used straight from the freezer.
- Care must be taken using hot water and any glass equipment.
- Safety issues should be discussed with the children.

Discuss how we keep things and ourselves warm, including inventions such as the vacuum flask. A simple starter activity is to put hot water in two beakers, wrap one in a material (a woolly hat!) the other is left unwrapped. Ask the children to predict which will cool down first and why. Leave thermometers (or temperature probes) in each and see which one cools down first. A parallel activity can be set up at the same time using ice.

The concept cartoons *Ice pops* and *Snowman* could be used. A good opportunity to investigate the best material for keeping different things warm/cold. A variety of problems could make this investigation relevant to everyday situations or a story/poem stimulus could be used such as the story of Goldilocks. Different groups could investigate different aspects of the situation eg

(a) Find the material to wrap the bowls to keep the three bears porridge warm!

(b) Keep the frozen peas cold/the cakes warm for Red Riding Hood on the way to grandma.

(c) Keep the tea in the teapot warm and the boiled eggs warm for breakfast.

Younger children may need help planning this, but essentially the children are testing a variety of materials.

Use the same area of material to cover the beaker and keep the volume of water the same.

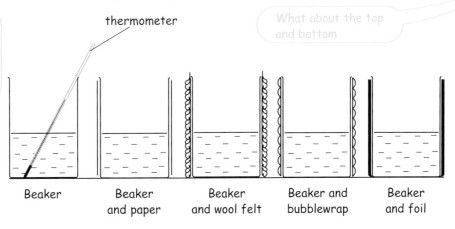

What about the top and bottom

Beaker | Beaker and paper | Beaker and wool felt | Beaker and bubblewrap | Beaker and foil

A chart to show the temperature of water after 15 minutes for a selection of materials

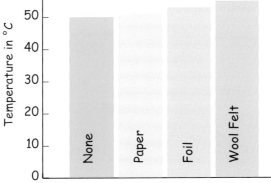

Use the same dimensions for the material and same volume of water.

Graphing package

INSULATION AND CONDUCTION

RS•C

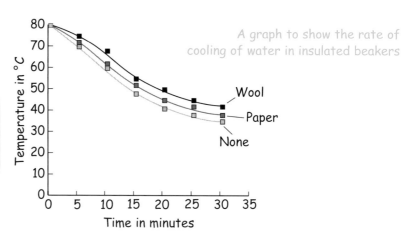

A graph to show the rate of cooling of water in insulated beakers

> Use different coloured pencils to plot the results from the different insulating materials.

Some materials are better thermal conductors than others

(a) At the simplest level children can stand spoons/rods made from plastic, metal, wood and glass in very hot water. If blobs of butter are put on the handle of the spoons, the good conductor melts it first. To eliminate the effect of the hot air rising from the container and melting the butter, stick the spoons through a cover or lid. Older or more able children could compare rods of different metals if you have them.

> Try to find spoons of a similar size.

> Do not put the butter too far up the spoon.

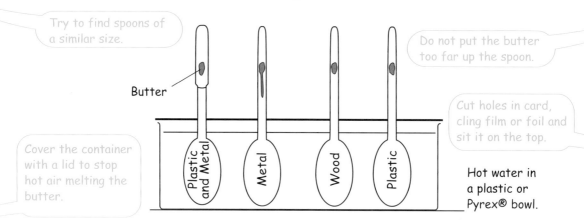

> Cut holes in card, cling film or foil and sit it on the top.

> Cover the container with a lid to stop hot air melting the butter.

Hot water in a plastic or Pyrex® bowl.

(b) For a quantitative result for older children use mugs of the same size but different materials. Tape thermometers to the outsides of plastic, metal and ceramic mugs and fill them with hot water. The temperature rises visibly at very different rates and could also be timed. What happens to the temperature of the water inside the mugs?

> Try to get mugs of similar thickness.

> Tape the thermometer to the mug with the bulb touching.

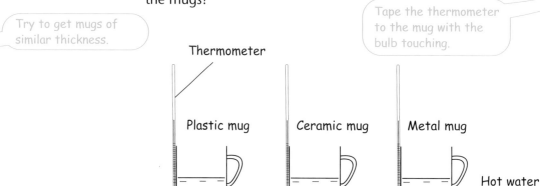

(c) **Investigate** 'All metals are good thermal conductors'. Compare metals only.

INSULATION AND CONDUCTION

Some materials are better electrical conductors than others

Metals are good conductors of electricity, most other materials are not

There is an opportunity here to discuss conservation of energy in terms of insulating our homes and not wasting heat. Older children may also discuss why it is warmer on a cloudy night and the 'greenhouse effect'. This is a good opportunity to use books and CD-ROMs to find out more about this.

A simple classification investigation which children could plan themselves to see which materials will conduct electricity. The children need knowledge of a simple circuit in order to do this and first test it out by touching the crocodile clips together. Then they try completing the circuit with a variety of different materials bridging the crocodile clips. They can construct a chart of results. Through discussion get the children to make generalisations about the fact that the metals conduct and that most other materials do not.

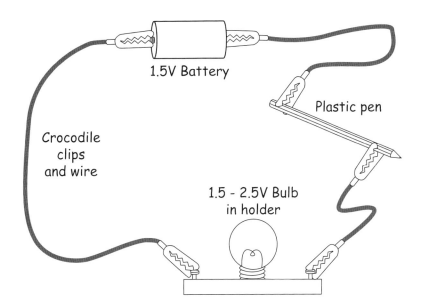

Word processing

Object	Material	Conducts Electricity	
		Prediction	Results
Pen	Plastic	No	No

(a) Get the children to test a greater variety of metals than they did for the previous activity.

(b) A good time to discuss general 'safety of electricity' issues with children as in the home, railway lines, overhead cables and pylons. Electricity providers are usually very willing to come and give a talk and show a video.

INSULATION AND CONDUCTION

star* Poetry
by Michael Rosen

Woolly Hats This poem can be used with any age group especially younger children discussing the properties of the materials of clothes that we wear.

Wooden Spoon This poem may need more explanation with a discussion about cooking utensils and would be a good springboard for the conduction activity.

Tank Jacket This poem may need a lot more discussion with younger children but is very appropriate for an investigation into the properties of thermal insulation for older children.

Simplicity of Electricity This is about electrical conduction and may be more appropriate as part of a topic on electricity.

Woolly Hats

In winter
when we go for walks we take hot drinks
in flasks
and we bury them deep
in our bags
wrapped up in woolly hats

In summer
when we go for walks
we take cold drinks
in flasks
and we bury them deep
in our bags
wrapped up in woolly hats

Wooden Spoon

He wouldn't use
a wooden spoon

but a wooden spoon
would do.

With it being so hot
in the cooking pot

a wooden spoon
would keep cool.

Wouldn't you sooner
use a wooden spoon

if you knew
the pot was hot?

Maybe you would.
Maybe not.

Tank Jacket

My dad said
the new tank
in the cupboard
needs a jacket.

I thought
a jacket?
What does it need
a jacket for?
It's not going out.
It hasn't got arms.
It hasn't got anything
to put in pockets.

My dad said,
the new tank
in the cupboard
needs a jacket.

So he went out
and brought it back,
and put it on the tank.

It didn't have sleeves.

It didn't have pockets.

The tank's not going out.

What a waste of money.

Simplicity of Electricity

The simplicity
of electricity
is that it never tires
of going down wires.

The invention of plugs
was to stop mugs
who touch wires
becoming electric fires.

The simplicity
of electricity
is that it seems to know
it cannot flow
through plastic.

Fantastic!

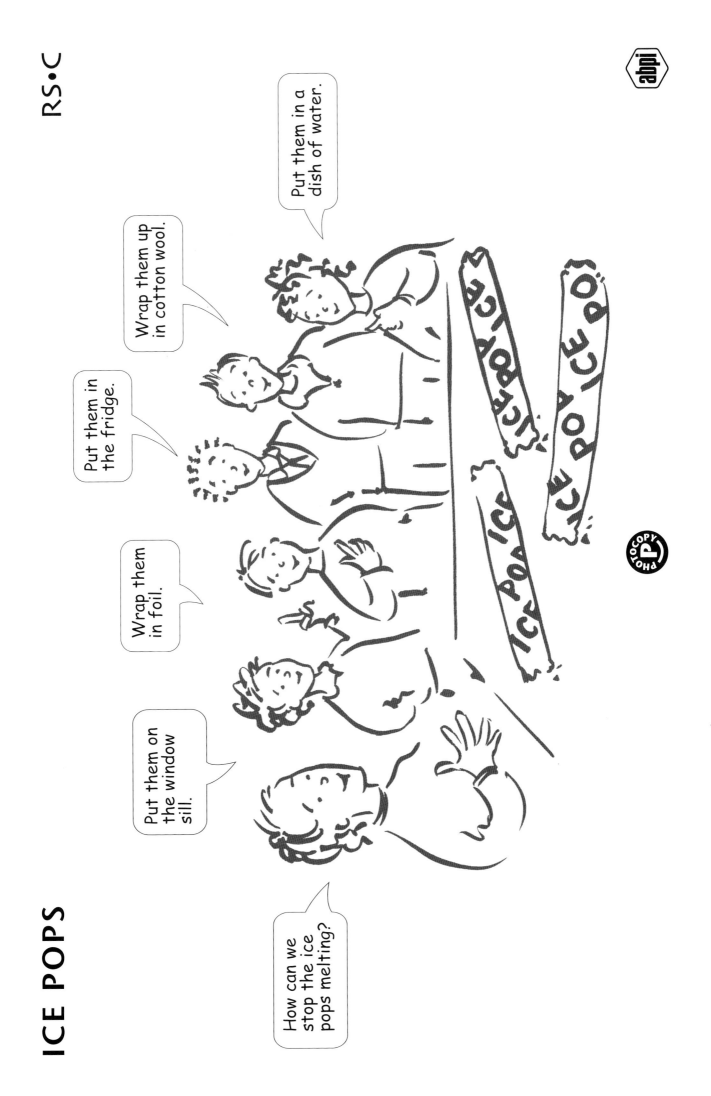

CONCEPT CARTOONS

Snowman

The issue in this concept cartoon is whether the coat is an insulator or whether it actually generates heat. Some children may believe that warm clothes make you warmer by making more heat, and they will expect the coat to generate heat and melt the snowman faster. However others will realise that the coat is simply an insulator which will tend to keep heat away from the snowman and prevent it from melting quickly. The situation shown in the concept cartoon can be investigated using real snow. Alternatively it can be modelled with ice inside a coat, glove or sock; the top half of a plastic mineral water bottle, filled with water and frozen, will make a good model snowman. The thickness, colour and nature of the material that the coat is made from can also be investigated.

Ice pops

All of the predictions in this concept cartoon can be directly investigated by the children. Some of them are likely to think that aluminium foil is an insulator; that cotton wool makes things warmer; that water will keep the ice pop cold; and that things will stay frozen inside a refrigerator. In each case they will be surprised by their observations! This can lead on to a whole series of follow up investigations on conductors, insulators and heat transfer.

this page has been intentionally left blank for your notes

ROCKS AND SOILS

Science background for teachers

VOCABULARY

Names of different rocks and soils; slate, marble, chalk, granite, sand, sandstone, clay, sedimentary, igneous, metamorphic (for older pupils)
Words to describe different types of rock; rock, stone, boulder, cobble, pebble, gravel, sand, clay, texture, rough, smooth, hard, soft, permeable, impermeable, porous, non-porous, porosity, absorbent
Names of physical processes; weathering, erosion

Note the treatment given here is more detailed than usual to aid teachers understanding although some simplification is necessary to prevent over complication.

Deep within the earth, hot, molten material called magma is formed. At times, this is forced out onto the Earth's surface when volcanoes erupt, where it cools quickly. Sometimes it is forced into the surrounding rock underground, where it cools slowly. Rocks formed in this way are called igneous after 'ignis,' the Latin word for fire. As molten magma cools slowly, minerals within it separate out as crystals, and the slower it cools, the larger the crystals grow. Granite is an example of this. Faster cooling magma from volcanoes forms rocks with smaller crystals, such as basalt. Both types are hard rocks often used in building. Also, frothy, gas-filled lava from a volcano can cool very quickly to form a very light rock, called pumice. The formation of the two other rock types, sedimentary and metamorphic, can be described in terms of what happens to these igneous rocks.

As pieces of rock are broken off into small particles by the action of weathering eg the freezing and thawing of water in cracks, they are transported by water or wind and finally deposited in layers in the sea as sediment. Physical changes may then take place, such as compression by the weight of later deposits. Chemical changes may also take place such as cementation, when water from the ground seeps through sand, depositing minerals between the grains and forming a natural cement. The cemented particles become sedimentary rock such as sandstone. Sedimentary rocks which mostly consist of particles larger than 2 mm, are called conglomerates if the particles have rounded shapes, and breccias if they are sharp and pointed. Other sedimentary rocks may also be made up of calcium carbonate, which is the main constituent of the shells of marine animals. When they die, their soft parts decay leaving the shells which eventually form limestone. Chalk is one important variety of limestone. Sedimentary rocks may also be formed from plant remains as coal. Sedimentary rocks may be softer and easier to cut than igneous rocks but also more easily eroded and affected by weathering.

Sedimentary rocks may be further changed when movement in the Earth's crust compresses them or pushes them deeper into the crust where they are affected by heat and/or pressure. These rocks are called metamorphic, from the Greek word for 'change'. Shale (fine silt and mud) changes to slate, limestone (mostly tiny animals) to marble, sandstone to quartzite and coal to coke.

The whole process can be summarised in the Rock Cycle (see diagram overpage).

ROCKS AND SOILS

The Rock Cycle

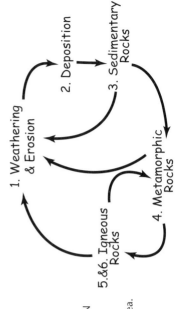

1. Weathering and Erosion
WEATHERING attacks rocks in many ways. Examples include acid rain, freeze-thaw, and plant roots. EROSION involves the transport and further breakdown of the weathered rock material by wind, ice, rivers and the sea.

2. Deposition
DEPOSITION of eroded mud, sand and pebbles often takes place in lowland areas. Most of this SEDIMENT is eventually deposited in the sea.

3. Sedimentary Rocks
As layers of sediment build up the lower ones become compacted under their weight and water is squeezed out. In time minerals dissolved in the water may be precipitated between the grains cementing them together, forming SEDIMENTARY ROCKS. Pebbles become CONGLOMERATE, sand becomes SANDSTONE, mud becomes MUDSTONE and shells become LIMESTONE.

(Over millions of years forces in the Earth's crust affect the sedimentary rocks. They may be squeezed during the collision of continental plates to form folds and be pushed above sea level where they may be attacked by weathering and erosion.)

4. Metamorphic Rocks
During plate collision rocks of any kind may be pushed to great depths where the pressures and temperatures are high enough to change them. The original layering and fossils are usually destroyed during recrystallisation but the rocks do NOT melt. Examples of changed or METAMORPHIC ROCKS include MARBLE (from limestone), SLATE (from mudstone) and QUARTZITE (from sandstone).

(After millions of years of uplift and erosion of the overlying rocks these metamorphic rocks may themselves be exposed at the Earth's surface and be attacked by weathering and erosion.)

5. Deep Igneous Rocks
At very high temperatures some minerals in the metamorphic rocks begin to melt. This molten rock, called MAGMA, may rise in the crust, then cool to form IGNEOUS ROCKS. Some cools slowly at depth, forming rocks with large crystals, such as GRANITE.

(After millions of years of uplift and erosion granite may be exposed at the Earth's surface and be attacked by weathering and erosion.)

6. Volcanic Igneous Rocks
Some magma may reach the surface of the crust and erupt as LAVA from VOLCANOES. BASALT is a typical lava and since it cools quickly it is made of small crystals. Frothy lava forms PUMICE and explosive volcanoes produce ASH. Volcanoes are built up from many layers of lava, ash and pumice. They are soon attacked by weather and erosion.

Reproduced by kind permission of John Reynolds ESTA

Rocks are made up of minerals, which may exist in many different forms. The crystalline forms can sometimes be cut and polished, for example the form of carbon known as diamond. Sometimes the minerals contain useful amounts of metals and are known as ores. Some minerals are changed during weathering and form other materials, eg the minerals in granite break down very slowly to form clay, sand and the oxides of aluminium and iron.

Many materials from the earth are used for building.

Clay has been used for centuries as a building material. On its own it is not very strong, but strong and cheap materials can be made from clay by heating it and mixing with other materials.

Bricks are made from clay heated to a very high temperature.

Cement is clay and limestone rock heated to a high temperature.

Concrete is cement, sand, stones and water mixed together.

Fredrich Mohs used rock minerals as 'scratchers', minerals that would either mark other minerals or would be marked by scratching. Using this test he then scaled rocks according to their hardness. So according to his scale, talc is the softest and will not scratch any other, diamond is the hardest, will scratch all the others and not be marked by any other.

Internet

Mohs Hardness Scale

(1) Talc	(6) Feldspar
(2) Gypsum	(7) Quartz
(3) Calcite	(8) Topaz
(4) Fluorspar	(9) Corundum
(5) Apatite	(10) Diamond

The fragments in sedimentary rocks are classified by size using familiar names.

Name	Diameter
Boulder	greater than 256 mm
Cobble	64-256 mm
Pebble	4-64 mm
Gravel	2-4 mm
Sand	1/16-2 mm
Silt	1/256-1/16 mm
Clay or mud	less than 1/256 mm

Soil is made up of dead plant and animal material mixed together with mineral particles formed by the weathering of rock. Micro-organisms in the soil cause the decay of the plant and animal materials which releases acids and nutrients into the soil. Soil formation begins as a result of the action of colonisation on bare rock of lichens, mosses and then small plants and animals, a slow process. Soil formation is faster when rock sediments are colonised, dead plant and animal material decays and mixes with mineral particles from the weathered rock.

Weathering is the process taking place when gases and water in the atmosphere combine with surface water and solar radiation to break up surface rocks. There are three main types of weathering:

- Mechanical weathering (physical) caused by, for example, rocks heating up, expanding during the day and rapidly cooling at night.
- Chemical weathering, the breakdown of rocks by chemical reaction. An example is the breaking down of limestone with rainwater.
- Biological weathering caused by plant roots stressing the rocks as they grow.

Erosion is frequently confused with weathering and is the breaking down of rocks by movement of rivers, ice, sea or wind.

Identification of Rocks

Rocks can often be identified using keys. Four rather complex examples are given overleaf for teachers to modify depending on the abilities of their children. A simple version is shown on page 46.

SKILLS

- Careful observation with a magnifying glass.
- Classifying according to given criteria.
- Carrying out repeated classification tests fairly.
- Recording results in chart form.
- Using a simple key.

ROCK IDENTIFICATION KEY 1 RS·C

Start here

Made from grains which may include fragments of fossils stuck (cemented) together → **YES** → This is SEDIMENTARY go to KEY 2

↓ NO

Made from crystals locked together

↓ NO / ↓ YES

Light and dark coloured crystals mixed together and with no obvious bands → **YES** → This is IGNEOUS go to KEY 3

↓ NO

Crystals are separated into bands → **YES** → This is METAMORPHIC go to KEY 4

↓ NO

Crystals are white and sugary → **YES** → (This is METAMORPHIC go to KEY 4)

↓ NO

START AGAIN!

This is probably a rock made of microscopic crystals or grains difficult to make a decision about → **YES** → Try all the keys, begin with KEY 2

A hand lens or magnifying glass is useful for these rock identification activities

These grains are stuck together.

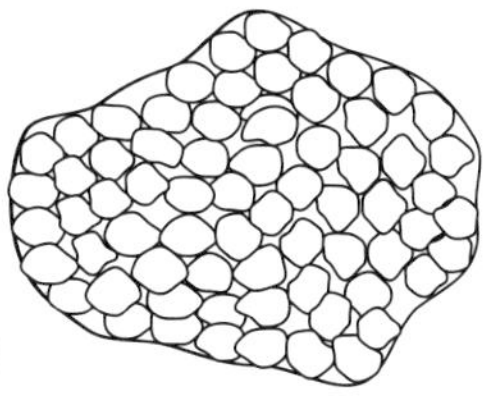

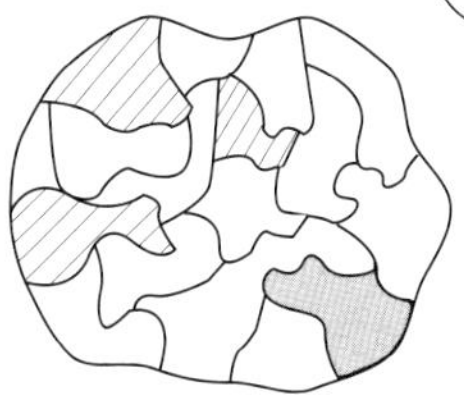

These are crystals locked together.

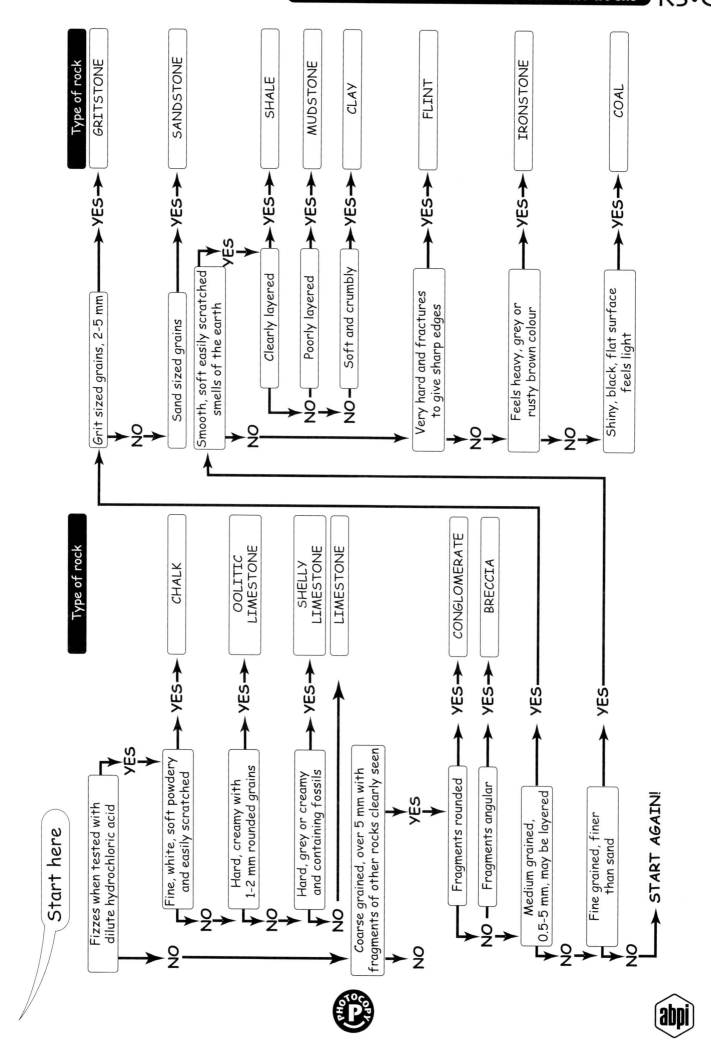

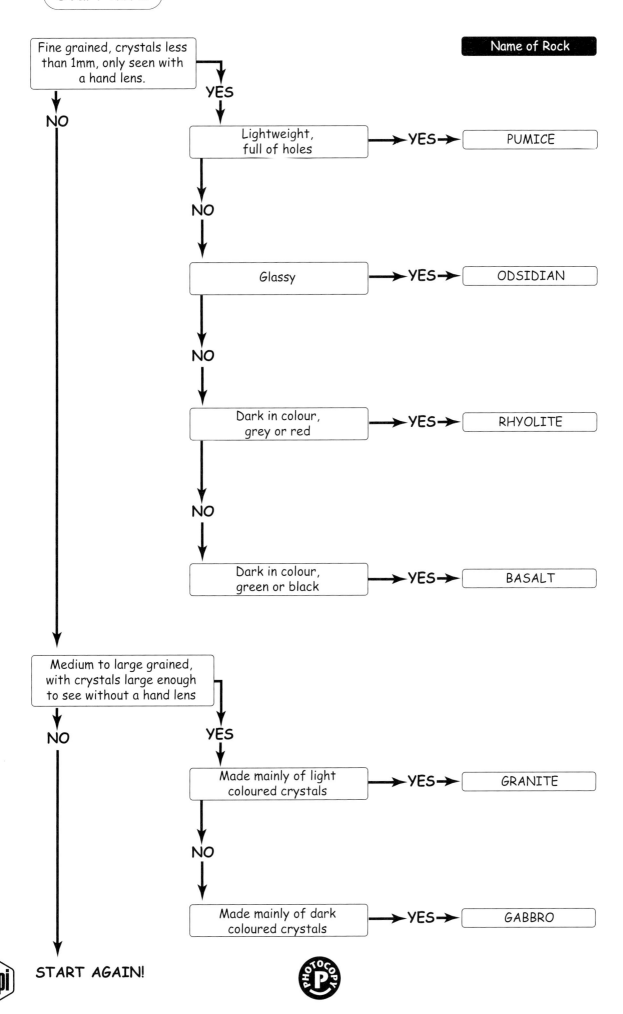

ROCK IDENTIFICATION KEY 4 – METAMORPHIC ROCKS

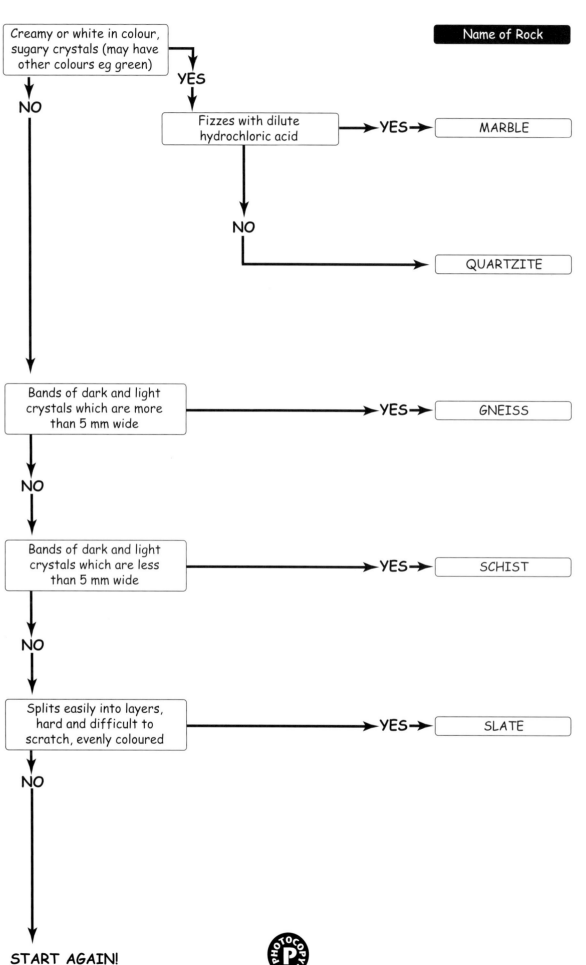

Key ideas and activities

Rocks are naturally occurring and used for a variety of purposes

(a) Make a collection of rocks and discuss other words used for rocks eg pebble, stone, boulder.

(b) Brainstorm the names of different types of rocks that the children may know. At this point it may be necessary to distinguish between natural rock and manufactured materials such as concrete that they call rock.

(c) Identify where rock is used in the environment, by looking at some pictures or photographs, or preferably going outside on a local trip into a churchyard, town or school grounds.

(d) Stone trail. Identify and chart the more common stones/rocks/bricks, include classifications such as natural or man-made, colour, where it was seen, use, age, damage.

(e) If you are near the sea, collect things from the beach; rocks, pebbles, shells, driftwood, seaweed and glass. Discuss the natural and man-made items. The collection can be used later to discuss how the sea and weather has affected them.

Rocks maybe grouped according to observable features

(a) Carefully observe rocks with a magnifying glass, draw and describe texture and colour.

(b) Let the children group the rocks according to their own criteria.

The children could then be shown the characteristics of six common rock types.

A simple rock identification based on six common rock types

Internet

(1)	Large crystals different colours	GRANITE
(2)	Sand grains, rusty colour.	SANDSTONE
(3)	Dark grey, splits into layers.	SLATE
(4)	Fossil shells, pale colour	LIMESTONE
(5)	Small, mostly black, crystals.	BASALT
(6)	White, sugary crystals.	MARBLE

(Reproduced by kind permission of John Reynolds ESTA.)

Rocks maybe grouped according to their properties

Discuss the importance of various properties of rocks (eg hardness) with the children in terms of the weathering and erosion that may take place to rocks that may be used for example in building. They may see it for themselves in the local environment on gravestones, walls and stone buildings.

ROCKS AND SOILS

Word processing

Set up a series of activities that can be done to test properties:

(a) **Hardness** Scratch test – Make a 'hardness' chart by carrying out a scratch test. Make only one scratch and if for example marble leaves a clear mark on the sandstone, then the marble is harder than the sandstone. Harder 'rocks' will scratch softer 'rocks'. Older children could scale the rocks 1-10, as in 'Mohs' scale. Other materials can also be tested and scaled accordingly.

Scratcher / Material	Chalk	Slate	Flint	Sandstone	Marble	Coin	Plastic counter
Chalk							
Slate							
Flint							
Sandstone							
Marble							
Coin							
Plastic counter							

(b) **Permeability** Using a dropper and water, see what happens when water is dropped onto different rocks. Some rocks will soak up the water, others will allow the water to sit on the surface. A simple observation test for younger children.

Is water absorbed by the rock, or does it sit on the surface?

Safety!
- Goggles should be worn when shaking or scratching rocks, or using acid.
- After handling rocks and soils hands must be washed or use plastic gloves.
- Soil samples should be checked for broken glass and animal faeces.

(c) **Porosity** The above test can be quantified by older children. Weigh the rocks dry, then immerse them in water. Bubbles can be seen as air from the pores in the rocks comes out. Leave for about 5 minutes. Which rock absorbs most water? Pat the rocks dry and weigh them again. The most porous rocks will have the largest weight gain.

Investigate which is the most porous brick? This test can be done with different bricks if you can acquire them. Different bricks are used for different parts of a construction. Hard, non-porous bricks must be used for the damp-proof course of a house where the bricks are in contact with the soil. More porous bricks can be used for internal walls. The results of these tests can be put into a table and the relevance of them to the everyday use of rocks discussed. Rocks which do not wear easily may be used for buildings, but because they are hard, they are difficult to cut and therefore expensive to use.

Use bricks that are the same size but may be different colours.

What are the bubbles?

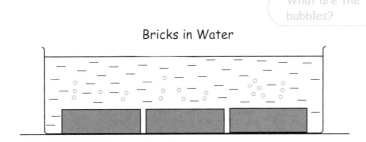

Bricks in Water

Use kitchen scales that go up to 3 kg.

(d) **Effect of acid** Using a dropping pipette and strong vinegar, see what happens when vinegar is dropped onto the rocks. Dilute hydrochloric acid (0.5 M where M indicates concentration in mol dm^{-3}) may also be used and works better on marble and calcite. (Try your local secondary school if this is not available from your primary supplier.) The calcium carbonate in limestone, calcite, chalk and marble will react with acid.

The chalk fizzes when you drop vinegar onto it. Use good quality, it's stronger.

Even if this is carried out as a demonstration, it is good practice to make the children wear goggles. Discuss 'Acid Rain' (acid from pollutants in the atmosphere dissolved in rain as it falls).

(e) **Magnetic** Are any of the rocks magnetic? (Lodestone/magnetite is and was used for compasses on ships. It could be borrowed from a local museum.)

(f) **Electrical conductors** Older children can test for conductivity. Although not magnetic, graphite is a conductor.

(g) **Using a key** Older children can be given a rock and by carrying out simple tests and observations may identify it using a key. Younger children may use a simpler key. An example is shown overleaf.

This key can be used to name rocks.

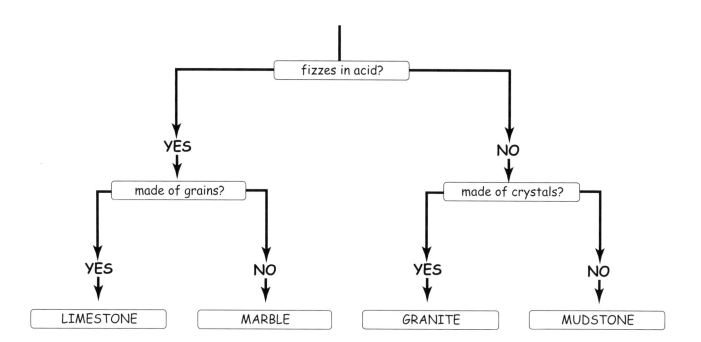

(a) Which **two** rocks fizz in acid?

_____ and _____

(b) Maria wants to make a statue.

It must **not** be damaged by acid rain.

It must be made of crystals.

Put a (ring) around the best rock to use.

   limestone   **marble**   **mudstone**   **granite**

ROCKS AND SOILS

Rocks breakdown to form the basis of soil

(a) **Rocks in a box (or plastic bottle)** Using rocks of similar size put them in a container with and without a small amount of water and get children to take turns shaking vigorously. The effect of weathering and erosion on different rocks can clearly be seen.

Erosion of the land A demonstration to show children how land may be eroded and the effect of de-forestation. Use a large, shallow, waterproof tray and put soil one end to represent the land. Leafy twigs, which represent trees and small, toy houses can be added. In the space the other end make a 'sea' from tap water. Using a watering can, 'rain' heavily on the land. The leaves catch most of the water and protect the land and houses. If the 'trees' are removed and it 'rains' again, the land and houses start to move. The land may slip into the 'sea'. This is a very effective demonstration.

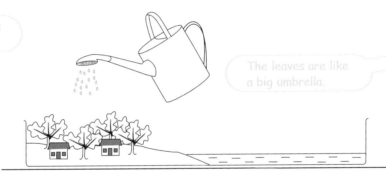

Now take the trees away.

The leaves are like a big umbrella.

Soil types to use: clay, peat, sandy, prepared compost, garden loam

You could use the concept cartoon **Sandcastles** here, but it may be more appropriate for Mixing and dissolving materials.

To look at the composition of and compare different types of soil

(a) Using a magnifying glass let the children look at different types of soil and describe the colour, feel and smell. Squeeze a handful of each soil; clay clumps together and sandy soil crumbles apart.

(b) **Sieving soil** Using graded sieve sizes, the soil can then be sieved into different sized particles and more closely observed for signs of animal and plants.

(c) Mix a sample of soil with water in a jar with a lid. Shake and leave to settle. This can be done with different types of soil to see the different layers settle out. Do they all take the same time to settle? Is there anything floating on the top, why do you think this is?

(d) As an extension activity, different samples of soil can be weighed, dried in an oven or over a radiator and re-weighed to see the water content. Investigate which soil holds most water.

(e) **Permeability** Using a funnel over a beaker, or a lemonade bottle cut in half with the top turned upside down over the bottom, tie muslin (or old nylon curtain) over the open stopper end to stop the soil falling through. Put soil into the funnel part and run water through it. Repeat with different soils. **Investigate** which soil is the most permeable.

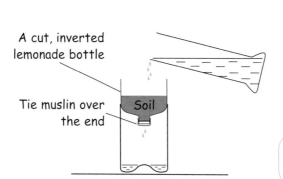

How much water is coming through the soil? Is there any on the surface?

Keep the amount of water and soil the same each time.

(g) **Air spaces** are important in soil for the things that are living and growing in it. Different soils have different amounts of air and the air spaces can be measured. Try this activity first. Using marbles, large stones, gravel, small stones and cubes fill beakers/identical jars. Now pour water from a measuring jug. Which jar needed most water to fill the air spaces? Repeat this with soils.

How much water will each beaker hold?

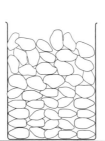 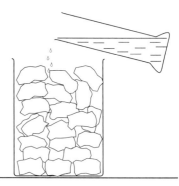

(h) **Acidity** Older children may test for this by adding distilled water to the soil, shaking then filtering the water off. This is then tested with Universal Indicator solution or papers. Commercial kits can be easily bought from garden centres or school suppliers. It can be a difficult test to get clear results, so it is a good idea to test the soil before letting the children do it.

Using secondary sources Videos, photographs and pictures are useful to show ways in which rocks are used and how weathering and erosion affect them and the natural landscape. The dangers of cliff tops and damaged buildings can be discussed. Children can access more information about volcanoes etc using CD-ROMs and the internet.

Internet

ROCKS AND SOILS · RS•C

by Michael Rosen

Yellow Rock As an introduction with younger children to talking about and looking at rocks generally. With older children, to introduce the origin of rocks and actions of volcanoes.

Yellow Rock

On holiday
on an island in Italy
we saw an old volcano.

We climbed up the side of it
and I picked up
a bit of yellow rock
and put it in my shirt pocket.

By the time
we got back
the stone
had made a hole in my pocket.

SANDCASTLES

CONCEPT CARTOONS

Sandcastles The distinction between melting and dissolving is a common area of confusion for children. They can clarify the meaning they attach to both of these terms by investigating the situation shown in the concept cartoon. A tray full of sand can be used to model the effect of the tide on sandcastles. Observation of other changes in materials, such as melting chocolate or dissolving sugar, will be a useful complement to their investigation.

this page has been intentionally left blank for your notes

SOLIDS, LIQUIDS AND GASES

Science background for teachers

VOCABULARY

Words to describe the different states of matter; solid, rigid, liquid, fluid, flow, pour, gas, gaseous state, melting, freezing
Names of solids, liquids and gases; ice, water, vinegar, oil, lemonade, oxygen, carbon dioxide, methane, helium

A solid has a definite shape that remains the same unless a force is acting upon it. The particles in a solid are rotating, vibrating or moving about a fixed position, close to each other. A solid normally occupies a slightly smaller space than the liquid, (except for ice, which takes up more space than liquid water). Some solids are made up of small particles eg sand. They can be poured like a liquid but the shape can be changed within the container, for example, when flour is put into a bowl, a well can be made in the middle.

A liquid has no fixed shape but a fixed volume and takes on the shape of its container. The molecules in a liquid move more and have more energy than particles in a solid but still remain in close contact with each other. Children often only think of water when talking about liquids, so it is worth beginning by brainstorming all the liquids that they know. Different liquids behave in different ways, some move more easily than others – they are less viscous.

A gas has no fixed shape or volume and will always spread out to fill the container it is in. The particles have a lot of energy, moving around in a random way, hitting other particles and the walls of the container.

All matter is made up of particles which have energy and move, the more energy they have the more movement there is. When a solid is heated it gains energy, the particles move more and it changes to a liquid state. If even more energy is supplied, more movement occurs and it changes to a gaseous state. If a gas is heated it gains more energy and takes up more space (expands). The concept of gases is difficult for young children. Older children can discuss gases with which they should be familiar as suggested below.

Methane is natural gas, a by-product of oil made under the sea from the decomposition of small plants and animals. It is the gas that is used as a fuel in our homes.

Steam is the name given to the gaseous state of water at or above 100 °C and is not visible. Water vapour is the gaseous state of water below 100 °C and is visible as tiny water droplets.

Helium is a gas that is completely non-reactive and has a low density which is why it is used as a lifting gas in fun balloons and airships.

Carbon dioxide is easily produced in the classroom. It is non-toxic, soluble in water and forms 'dry ice' at low temperatures. It is a product of respiration by plants and animals and used by plants to photosynthesise.

SKILLS
- Recognising and carrying out a repeat procedure fair test.
- Choosing and using equipment accurately – reading a thermometer and stopwatch.
- Working cooperatively in a group.
- Careful observation and recording.

Key ideas and activities

SOLIDS, LIQUIDS AND GASES

To know that materials can be classified as solids, liquids and gases and recognise the difference

Exploring solids and liquids can begin for children at an early stage, the concept of gases is more complex and probably better left until further into primary. The ideas and activities associated with solids, liquids and gases are obviously linked closely with the concept of change of state and heating and cooling materials which will also be dealt with on pages 81-85.

Solid, liquid and gas Have three identical balloons, the day before fill one with water and freeze. Fill one with water, and blow one up. The best way to fill a balloon with water is to put the end over the tap, turn the tap on and support the balloon underneath as it fills. Discuss the solid, liquid and gas.

Measure the circumference of the balloon with water before and after you freeze it.

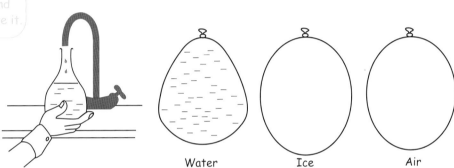

Water   Ice   Air

Using the balloons pictured above, undo/cut the top off the 3 balloons. Peel the balloon off the ice. Pour the water one into another container and let down the blown up one. Discuss the behaviour of each one.

Some solids have air trapped inside them Look at the spaces in a sponge and pumice stone closely with a magnifying glass. What do you think is in the spaces? Immerse the sponge in water and squeeze, what are the bubbles? (Also refer to the air/soils activity in the Rocks and Soils section.)

SOLIDS

Database

(a) **Looking at solids** Make a collection of a variety of solids to draw and discuss. Use this as an opportunity to develop descriptive language. Include solids of different 'hardness' such as a piece of wood and a sponge and pouring solids such as salt.

(b) **Some solids have similar properties to liquids, they can be poured** Using salt, sand, rice, lentils, etc, let the children explore the flowing behaviour of solids which are finely divided. Discuss the similarities and differences with liquids.

Use the concept cartoon *Is it a solid?*

SOLIDS, LIQUIDS AND GASES

LIQUIDS

(a) **There are liquids other than water** Draw, discuss and collect a variety of liquids and with older children discuss the viscosity or thickness of the liquid.

(b) **Race the liquids** Investigate which liquid moves the fastest. Choose a variety of different liquids eg water, syrup, tomato sauce, vegetable oil, glycerine, cream and put a measured spoonful of each at the top of a tray. Tip the tray and lean it against a brick allowing the liquids to run down the tray.

Spreadsheet

Which is the fastest? Older children could time the activity.
What happens if you cool all the liquids first in a refrigerator?
Compare different temperatures of the same liquid eg tomato sauce that has and has not been in the refrigerator.

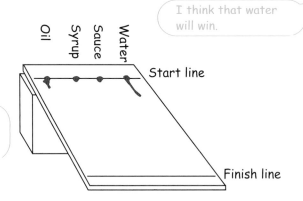

(c) **Conservation of volume of liquids with change of shape**
Using a given volume of water let the children pour it into a variety of different shaped containers and draw and observe the change of shape. Ask them to predict each time how far up the container they think it will go. Repeat the task with different volumes, as this gives younger children practice at measuring volume.

Modelling

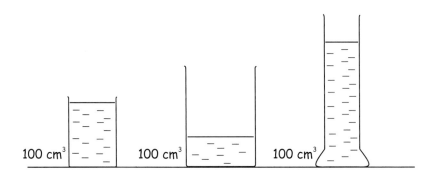

GASES (a) **Air is a gas, you can feel it and it has weight** Blow up a balloon and then let it down, feel the air escaping.
If you have a digital balance, weigh an inflated and deflated balloon. Another way to show that air has weight is to balance an inflated balloon against a deflated one. Use a piece of lightweight dowel and thread and balance it first with two deflated balloons. Then blow one up and retie, the inflated one just tips the balance.

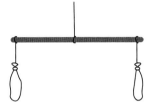

Balance the balloons on thin dowel

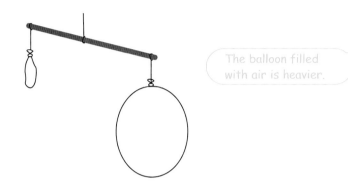

The balloon filled with air is heavier.

(b) **Warm air expands and rises** Put a balloon on the top of a small plastic bottle. Stand the bottle in a bowl of warm water and observe what happens. Why? Now take the bottle out of the water. What happens? Discuss hot air balloons with the children.

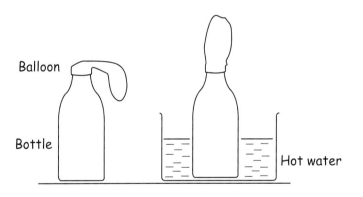

Hang a paper 'snake' over warm air and watch it turn.

Investigate some variables, eg temperature, does hotter water blow the balloon up more or for a longer time? Does the size of the bottle make any difference? Make snake shapes by cutting a continuous spiral out of paper and hang them on string on a hanger over a warm radiator, what happens and why?

(c) **Carbon dioxide** There are a variety of ways to make carbon dioxide. The one chosen may depend on the cross-curricular context of the lesson. In each case the gas is made by a chemical reaction and can be collected in a balloon that fits over the top of the bottle.

SOLIDS, LIQUIDS AND GASES

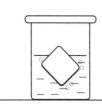

- Alka-Seltzer tablet and water. This is fast reacting and explosive and will blow the top off a film canister! The carbon dioxide can also be collected in a balloon on top of a bottle.

- Sodium hydrogencarbonate (bicarbonate) and vinegar, produces lots of 'froth'.

- Sodium hydrogencarbonate (bicarbonate) and lemon juice, which is an endothermic reaction. (The temperature of reaction mixture will decrease as heat from the surroundings is used up.)

- Baking powder and hot water. (The reaction that often occurs in cake making.)

- Dried or fresh yeast, sugar and warm water. This reaction takes 5-10 min. It is used in bread and pizza baking.

- Lemonade 'fizz' is carbon dioxide. Drop currants in new, fizzy lemonade and watch them dance up and down as the gas lifts them to the surface, pops, escapes and allows them to drop to the bottom.

Safety!
- Care when handling any glassware.
- Spirit thermometers should be used.
- Care when handling hot water.

Investigate the variables in any of the carbon dioxide reactions:
Temperature of the water in the yeast activity, try cold through to very hot.
The amount of sugar, does it affect the speed of the reaction?

Does the balloon blow up twice as big with two Alka-Seltzer tablets?

(d) Helium Buy a helium balloon and allow it to sit on the ceiling of the classroom until it falls. Why does it float? Why does it fall?

(e) Smells in the air! Air fresheners, perfume, perfume oil burners, food and cooking smells, why can we smell them in different parts of the room and the house? The particles that cause the smell are gaseous and reach our noses. You can demonstrate this for children and discuss what is happening.

SOLIDS, LIQUIDS AND GASES

by Michael Rosen

Steamy Shower Water changing to water vapour and back again.

Yellow Door The gas that is used in our homes.

Floating Balloon What is the gas in the 'floating balloon' and why does it end up on the ground?

Drizzy Fink The 'fizz' in lemonade is carbon dioxide.

Pouring You cannot pour a teapot but you can pour from one!

Steamy Shower

I love a dreamy, steamy shower
hanging about
for over an hour
just before bed
getting hot and red
in the steam
standing there
with time to dream
water-running-over-me feeling
drips dripping off the ceiling
mum says it's my fault it's peeling
nothing can beat
the hot wet heat
nothing wetter nothing better
I love a
dreamy steamy streamy shower

Yellow Door

I've often wondered where gas comes from.

Now I know.

I was walking down the street
and I looked down
at the pavement
and I saw
a little yellow door.

I thought:
why is there
a little yellow door in
the middle of the
pavement?

And then I saw it.

On the door
it said:
'GAS'.

Now I know where
gas comes from.

SOLIDS, LIQUIDS AND GASES

Floating Balloon

My balloon from the
fair hangs in the air
nosing the ceiling.
It's string hangs down
like a tail to the floor

I lie in bed watching
it tremble and it
quivers when I give a
blow.

I dream of peaches
that float round trees

But in the morning
my balloon from the
fair squats on the
floor its tail snaking
over the carpet

I get out of bed
watching it roll and it
bounces when I give
it a kick.

Drizzy Fink

Hail! Hail!
I come from another galaxy
I have been learning English
I find some of your words
very hard to say.
I will now try to talk about
your fizzy drinks.
I like your fizzy drinks.
I think I will drink lots
of fizzy drinks and
collect lots of bottles
and lots of tottle bops
    er
bopple tots
    topple stobs
    pobble lots
    no
    stottle pobs
    tobble spots
    lottle slobs
    lobble slops
Please
can you help me with this?
And please
can I have a drizzy fink?

Pouring

Whether it's hot,
or whether it's not
I don't see how
you pour a tea-pot.

I can see
you **can** pour tea,
but you surely cannot
pour a tea-pot.

IS IT A SOLID?

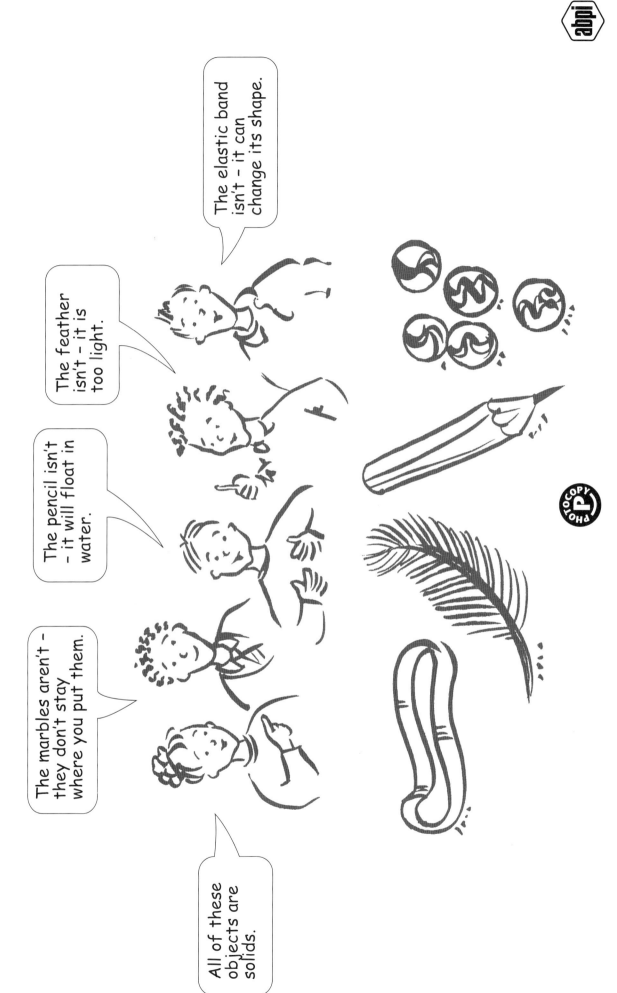

CONCEPT CARTOONS

Is it a solid? The children will have intuitive ideas about what they mean by a solid. However they will not find it easy to apply their ideas in a consistent way. They will find it difficult to separate the object from the material it is made from, and they will tend to associate properties such as heaviness and rigidity with solids. The concept cartoon provides an opportunity for them to rethink their definitions and to make more systematic judgements. Introducing more challenging materials such as sand or dough is probably best left until after their ideas about solids are reasonably well developed.

MIXING AND DISSOLVING MATERIALS

Science background for teachers

VOCABULARY

Mixture, solid, liquid, gas, solution, suspension, dissolve, factors, saturated, soluble, insoluble, permanent, reversible, separate, sieve, change, solvent

Mixing materials together forms a mixture where two or more substances are physically but not chemically combined and can be separated again by physical methods such as sieving, filtering, evaporating etc. This means that the change can be reversed. A new material is not formed as it is in a chemical reaction. Mixtures can be made up of;

- solid in solid (muesli)
- gas in solid (pumice stone)
- solid in gas (smoke)
- gas in gas (air – mostly made up of nitrogen and oxygen)
- liquid in gas (clouds, mist, aerosol)
- gas in liquid (oxygen in water)
- liquid in liquid (emulsion – milk)
- insoluble solid in liquid (suspension – muddy water)
- soluble solid in liquid (solution – salt water).

Mixing materials together may form a new substance but not a mixture, as with Plaster of Paris and water. A chemical reaction has taken place in which case a completely new substance has been formed and the original substances cannot easily be recovered. The word 'mixture' is not always used correctly in its scientific sense. A cake mixture for example is so until water is added and it is baked. A chemical change has then taken place and it is no longer a 'mixture' but a cake! A safe way of referring to such situations might be to call them a 'mix'. Activities involving baking mixtures have been left out of this section and are included in the section 'Heating and cooling materials'.

DISSOLVING

When the mixture is a solid in a liquid it will either produce a solution or a suspension. A solution is clear and will never settle out, a suspension is cloudy and will eventually settle out. If the solid (solute) dissolves in the liquid (solvent), a solution is formed. Although at primary level the solvent mostly used is water, older children need to know that there are others. These will dissolve other solids that may not dissolve in water eg nail varnish is dissolved by propanone (acetone). Various factors that affect dissolving will be familiar to children, such as stirring, temperature, time, amount of solute, amount of solvent and can be equated with their everyday life such as stirring sugar in a cup of tea.

Sometimes a solid dissolves in a liquid to produce a solution, but not of the original substances, because the change is a chemical one, for example when Alka-Seltzer dissolves in water.

For younger children, it is sufficient to know that a change of some sort has taken place, older children can explore whether it is permanent or reversible.

SATURATED SOLUTIONS

Only a finite amount of solid will dissolve in a liquid and this is dependent on the solid and the temperature. When no more solid will dissolve the solution is saturated, but generally more will dissolve in hot than in cold liquid. So if a saturated solution of sugar is made using hot water, as it cools some of the sugar comes out of solution and reforms making crystals. This principle is used to grow crystals. Crystals can also, of course, be made by leaving a salt or sugar solution and allowing the water to evaporate leaving salt or sugar crystals behind.

Children often confuse melting and dissolving and this may be a point of discussion. Melting requires heat and dissolving requires a solvent. In making a jelly both are happening when it is added to hot water.

SKILLS

- Using a thermometer and stopwatch.
- Working cooperatively.
- Choosing and using apparatus carefully.
- Observing and recording accurately.
- Measuring accurately.

Key ideas and activities

Mixtures have more than one thing in them (some may be separated by sieving)

(a) Look at and discuss with the children obvious mixtures such as Dolly Mixtures, a jar of mincemeat, mixed vegetables, fruit salad, muesli. Get the children to closely observe and list the ingredients in the mixtures. Ask the question, 'Can they be separated?'

(b) It may be appropriate at this stage also to do a sieving activity to separate the various components of a mixture such as peas and rice and allowing the children to choose their own equipment or to make up their own mixture. Explain to them that to call it a 'mixture' they must be able to separate it again. This would make a good combined activity for younger children especially if some sieving is also done with soils (see Rocks and soils section). Investigate how to separate sand and peas, or stones and sand.

A food Design Technology activity where the children make simple 'mixtures' such as icing sugar and coconut to make 'coconut ice', design their own fruit salad or make 'chocolate crispy' cakes using melted chocolate and 'Cornflakes' or 'Rice Krispies', can also be carried out.

Mixing materials can cause them to change

Making pencil crayons This is a directed activity where the children need to follow a recipe for making pencil crayons. They will need:
- Large, marble-sized ball of modelling clay
- Small spoon of paste, (cellulose wallpaper paste made up with water beforehand)
- Big spoon of powder paint or iron oxide (pottery supplier or local secondary school). The paint makes 'playground pencils', the oxide makes brown thick pencil leads that write on paper.

Mix the mixture outside the bag. After rolling out the 'leads', wash your hands.

Mix the ingredients in a polythene bag by squidging it from the outside, then take it out and mix it in the hand, it resembles crumbly plasticene. Roll out the 'leads', this makes about three 8 cm lengths and leave to dry overnight on paper towels. The 'leads' can be covered in fancy paper.

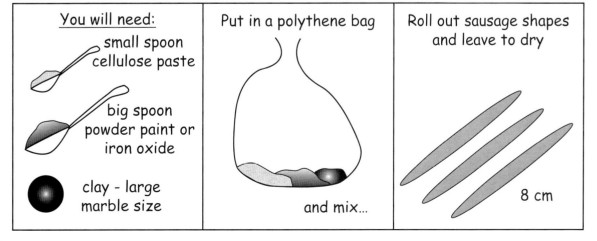

MIXING AND DISSOLVING MATERIALS

Some materials dissolve in water others just mix

(a) Present the class as a whole with a wide range of materials to mix so that there can be a comparative, plenary discussion. Each group of children can have a smaller selection to work with. Include soluble and insoluble materials and get the children to record what happens each time. Suggest they stir or shake the mixtures and be patient! This would be a good lead into work on dissolving.

Good examples are;

Water with sand, sugar, salt, Alka-Seltzer, oil, coffee, lemon juice, flour;

Vinegar with oil, lemon juice, bicarbonate of soda.

At this stage the amount of each ingredient mixed is not critical but is used in the case of solids in liquids, 1 teaspoon of solid to at least 100 cm^3 of water to avoid saturation. Keep the various mixtures for the discussion.

(b) The results of the activity could then be grouped into those materials that dissolve and those that do not, having discussed solubility first. In some of these mixings has anything new been made?

Word processing

You could use the concept cartoon on **Sandcastles** here.

Water and:	Prediction: will it dissolve?	Did it dissolve?	Describe any other changes that occurred
Sand			
Chalk			
Sugar			
Salt			
Alka-Seltzer			
Oil			
Flour			
Coffee			

The coffee dissolved and went brown.

The Alka-Seltzer dissolved and fizzed.

There are a variety of factors which affect dissolving

Discuss this with the children first and get them to suggest the factors that affect dissolving from their own experiences such as making a cup of tea. Get them to suggest a hypothesis, for example; 'I think that stirring affects dissolving' and then plan a way to prove it. This would make a good investigation or series of small investigations, eg Investigate Does stirring affect dissolving? This can have particular reference to fair testing due to the large number of variables that could be investigated. In this set of quantitative investigations, care should be taken if using salt, as impurities in the salt often leave the water cloudy after it has dissolved. Consequently, it is not always clear at what point the salt dissolves.

(a) **Stirring** Limit the number of soluble solids eg sugar which works very well, salt (test for cloudiness first) and water even if they are planning this themselves. They can time how long dissolving takes with and without continuous stirring and stirring once every minute. Suggest 1 teaspoon of solid, even then dissolving without stirring can take a long time.

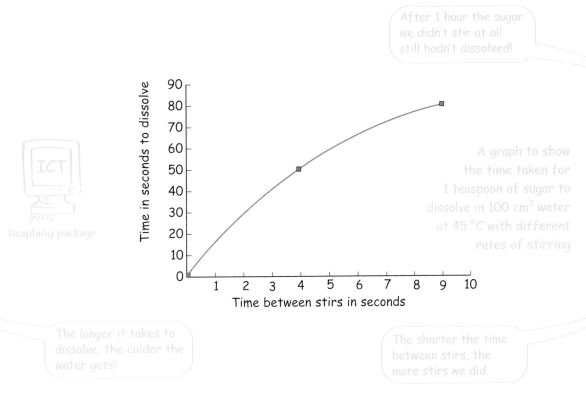

After 1 hour the sugar we didn't stir at all still hadn't dissolved!

A graph to show the time taken for 1 teaspoon of sugar to dissolve in 100 cm³ water at 45 °C with different rates of stirring

The longer it takes to dissolve, the colder the water gets!

The shorter the time between stirs, the more stirs we did.

(b) **Temperature** Repeat activity (a) using hot and cold water. Suggest one good stir at the beginning or continuous stirring, otherwise it can take a long time to dissolve the solid and the temperature of the water will drop.

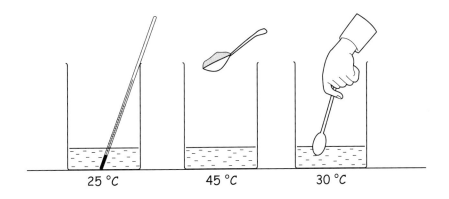

We used 100 cm³ of water each time and kept stirring until it all dissolved.

MIXING AND DISSOLVING MATERIALS

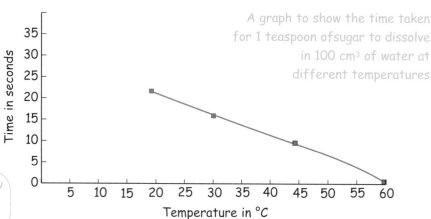

A graph to show the time taken for 1 teaspoon of sugar to dissolve in 100 cm³ of water at different temperatures

Can you predict how long it would take at 5 °C?

(c) **Size of the solid particles** For convenience of time, use warm water and stir. Present the children with one solid in different forms where the grains are different sizes eg sugar – granulated, castor, icing, sugar crystals. Let the children look at the solids with a magnifying glass first and predict which will dissolve the fastest. Brown sugar contains other substances, so is not the same as white sugar and not appropriate to use.

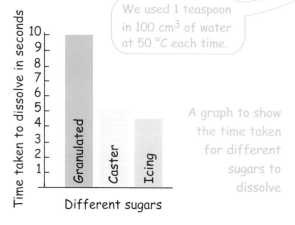

We looked at the size of the sugar grains with a magnifying glass. The granulated sugar had the biggest grains.

We used 1 teaspoon in 100 cm³ of water at 50 °C each time.

A graph to show the time taken for different sugars to dissolve

Safety!

- Care should be taken when handling glassware.
- Spirit thermometers should be used.
- Care should be taken when handling hot water.
- Remind children NOT to sample any of the ingredients – sugar is tempting.
- Some solvents are highly flammable and the room must be well ventilated.
- Polythene bags are potentially dangerous, use small ones and supervise closely.

(d) **Amount of solid** Again for time convenience, use warm water and stir, eg once every minute. Present the children with only a few soluble solids and ask them if the amount they put in the water affects the dissolving time.

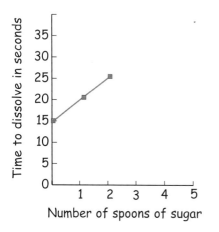

A graph to show how the dissolving time of sugar changes with the amount of sugar mixed with 100 cm³ of water at 55 °C.

Can you predict how long 4 teaspoons would take to dissolve?

(e) **Investigate The fastest way to make a jelly!** There are many variables to be considered, temperature and quantity of the water, size of the jelly cubes, with/without stirring and temperature for setting of the jelly. Before they start the children could make a list of all the factors that they have found out that speed up dissolving. An excellent planned investigation for older children, younger or less able children will need help with coping with all the variables.

The following activities would be suitable as extension activities for more able 11 year old children, as ideas for a Science 1 investigation or ideas for a Science Club.

Note – There is a limit to the mass of solid that can dissolve in a given amount of water and this is different for different solids.

Activity (d) could lead nicely into the following activity, especially if any of the children put so much solid in the first time that it did not dissolve! A good investigation for able 10-11 year olds; 'is there a limit to the amount of sugar that will dissolve in 100 cm^3 of water? Is it the same for salt?'

- Keep the volume and temperature of water constant and increase the amount of solid each time until a saturated solution is reached, (ie some solid is left). Record the number of spoonfuls that will dissolve.

- Repeat the above activity with a different solid. Is the saturation point the same?

- *Growing crystals* Great fun! This can be done as a demonstration or with a small group and the whole class can watch the progress over a few days. *Tip* – The best crystals are obtained using alum (aluminium potassium sulfate – available from any chemical supplier). Sugar and salt crystals are very difficult to grow and copper sulfate should not be used in the primary sector.

Experimental method: Make a saturated solution in a jar using hot water (add as much solid as you can dissolve). Decant the solution (pour off) into a clean jar leaving behind any undissolved solid. Using a pencil, or the lid of the jar with a hole through, suspend a thread into the solution. If a lid has not been used, cover the jar opening with taped paper, to control the rate of evaporation and prevent dust dropping in on the forming crystals. Treat with care and leave undisturbed. Crystals will grow on the thread after a day or so.

Substances that do not dissolve in water can be dissolved by other solvents

It is worth demonstrating with older children how alternative solvents can be used. For example use at least one other solvent such as propanone (acetone) to dissolve nail varnish or white spirit with paint or detergent with oil.

by Michael Rosen

Mayonnaise An emulsion made with eggs, vinegar and oil, which can be tricky to mix!

Making Jelly How do you make a jelly?

Summer Sand Some solids like sand do not dissolve.

Water bottles and tubes More sugar than salt will dissolve in a given volume of water.

Drizzy Fink Lemonade, carbon dioxide in water with lemon flavouring. See page 60.

Mayonnaise

A French friend of dad's
was making mayonnaise.

He poured
and mixed
and whisked
for hours
and he kept saying:
It is going to be marvellous,
it is going to be superb.

A woman walked past.

He cried out:
It is destroyed.
It is completely destroyed.

Destroyed? Said my dad.
Yes, he said,
the lady with the beard breathed on it.

Later, at home,
dad said,
I didn't see her breathe on it.
And I said,
I didn't see her beard.

Making Jelly

It's my job
to take the slab of jelly
and break it up into cubes.

It's mum's job
to pour on the boiling water
to melt the cubes.

It's my job
to stir it up
until there are no lumps left.

It's her job
to put the bowl
in the fridge to help it set.

It's my job
to eat it.

Summer Sand

The washing machine
rinses the summer holiday away
and when everything's dry
it all goes into
my chest of drawers,
with the clothes
I didn't take
That's it.

Summer over
till next year.
The beach, the surf,
the sun, the wind,
all washed away.

Weeks later
I am putting on a pair of socks
and there's the summer!
The beach, the surf,
the sun, the wind
in the sand
caught in the toe of a sock.

Water bottles and tubes

When my sister
is better
she won't have to lie in bed
between two upside-down
water bottles
with tubes going into her.

They must know my sister very well,
because one's got a little bit of salt in it
while the other's got
quite a lot sugar.

They could have done better
with that sugar one, though:
they should have made it
with chocolate milk shake.

Never mind.
When my sister
is better
she won't have to lie in bed
between two upside-down
water bottles
with tubes going into her,
and I'll make her
a big fat
mega-mega
chocolate milk-shake -
KER-PAM !!!

SANDCASTLES

CONCEPT CARTOONS

Sandcastles The distinction between melting and dissolving is a common area of confusion for children. They can clarify the meaning they attach to both of these terms by investigating the situation shown in the concept cartoon. A tray full of sand can be used to model the effect of the tide on sandcastles. Observation of other changes in materials, such as melting chocolate or dissolving sugar, will be a useful complement to their investigation.

this page has been intentionally left blank for your notes

HEATING AND COOLING MATERIALS

Science background for teachers

VOCABULARY

Solid, liquid, gas, change of state, physical change, chemical change, heating, cooling, melting, freezing, boiling, evaporation, hot, cold, condensation, steam, water vapour, reversible, permanent, irreversible, Celsius, thermometer

Heating and cooling materials can be associated with a change of state between solid, liquid and gas if the temperature changes are large enough. If the change is just physical then it can be reversed, as in melting chocolate or freezing water, but sometimes heating brings about a chemical change and is permanent as with heating (firing) clay or cooking eggs. Albumen (egg white) is mostly water but about 9 per cent protein and when heated the protein structure is changed permanently. For younger children a chemical change is very difficult to understand and best left to description only, 'what happens when we…?' For older children questions can be asked about the reversibility or not of the process and permanence of the change, and whether or not they think a new material has been formed. This increases their awareness of the different types of change being possible and helps to develop the idea of a chemical change.

In the solid state, the particles that make up a material are vibrating about a fixed point, packed closely together. As heat is applied, energy is gained and the particles vibrate more, reducing the forces holding them together. They move apart, breaking the solid structure and become a liquid. This is called melting. When further heat is applied, the particles move even faster, breaking away from each other and forming a gas or vapour (evaporation). When this happens to all the particles it is called boiling.

On cooling, the particles in the vapour slow down, come together and loosely bond, returning to the liquid state (condensation). Further cooling slows the movement of the particles even more and they become solid. If this happens at or near room temperature, as with chocolate or wax they are said to solidify. If this takes place at cold temperatures as with water it is known as freezing.

Children need to be aware of these changes in a variety of materials including water, which has an unusual property.

WATER

Most substances decrease in volume when cooled and increase in volume when heated. Water is no exception to this, except at temperatures between 0-4 °C, because when water melts, it contracts. At 0 °C, water is ice, which has an open honeycomb structure. This takes up more space than water in the liquid state and so there is an increase in volume. This is why pipes burst and the tops pop off frozen milk bottles. As the ice begins to melt between 0-4 °C, the honeycomb structure collapses and the volume decreases, so at 4 °C its volume is at a minimum and it is most dense. This also explains why ice floats on water and in ponds there is life under the ice. The ice at 0 °C is less dense than the water between 0-4 °C, therefore the ice floats and the water stays at the bottom of the pond. Above 4 °C, the water expands as it warms up.

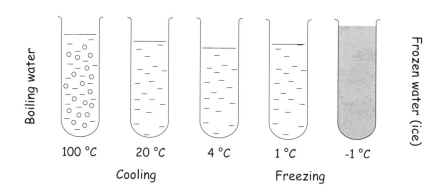

A diagram (exaggerated) to show changes in volume of water at different temperatures

If chemicals such as salt are added to ice, the freezing point is then lowered to about -10 °C. Salt on the surface of ice dissolves forming a salt solution with a lower freezing point than the ice on which it sits. This causes the ice to melt, more salt dissolves and so on. This process also causes the temperature to drop and is called an endothermic reaction.

When water is heated to 100 °C it boils and the liquid turns to a gas, which are the bubbles seen in boiling water. The gas escapes as invisible steam, the name given to the gaseous state of water at 100 °C or above. The visible cloud we often call 'steam' is actually tiny, condensed, water droplets that are light enough to remain suspended in the air as a mist or a cloud. Water can also exist as a gas in the air at a temperature below 100 °C, this is known as water vapour. When steam or water vapour get a lot cooler they condense back to the liquid state.

EVAPORATION

When liquids are left out and exposed to the air, they eventually 'dry up' or evaporate. This is due to the particles on the surface of the liquid being able to escape, turning to the gaseous state and moving away from the rest of the liquid. As the particles close to the surface move away, the liquid decreases until none is left. The process of a liquid changing to a gas below its boiling point is called evaporation.

TEMPERATURE

Temperature is a measure of how hot or cold something is and can be measured in a variety of ways. The most common piece of apparatus for measuring temperature in primary schools is the spirit thermometer with temperature measured on the Celsius scale. A thermometer is a thick glass tube with a very thin bore filled with liquid, which ends in a thin-walled reservoir. The liquid in the reservoir is very sensitive to temperature change and as the temperature increases, the liquid expands and rises up the narrow bore. 'Feeling' the temperature of something is not a very reliable measure due to the different thermal conducting properties of different materials. (See section on Thermal insulation.)

SKILLS

- Using a thermometer with care, reading the scale.
- Using a stopwatch accurately.
- Measuring volume and mass with care.
- Careful observation.
- Tabulating results and constructing graphs.
- Planning for a fair test, using equipment with care.
- Working cooperatively in a group.
- Following instructions.

Key ideas and activities

These begin with activities for younger children (4-7 year old students) and progress to older children (8-11 year old students).

Some materials change when they are heated, some changes are reversible and some are not

Word processing

(a) Put small quantities of different substances eg chocolate, wax, butter, margarine, ice cube, cheese, pasta, plasticene into small, plastic 'mousse' pots or tin-foil pastry cases, and stand them or float them on hot water. The children can observe the substance melting. Do they all melt, if not, why not? Now put them in cold water and watch them solidify.

(b) Children could investigate 'which melts the fastest?' To keep the test fair, compare the same quantities, eg equal cubes of margarine, butter and chocolate.

(c) **Different materials melt at different temperatures** Older, more able children can investigate the fact that materials melt at different temperatures. This may be apparent to them from the outcome of activity (a). At primary level with simple equipment, this activity is limited to a few materials the children can melt. However, in a practical way it introduces the concept of materials melting at different temperatures. It may also lead to a discussion about the temperatures needed for other materials such as metals to melt and whether or not all materials will melt.

If it doesn't melt over hot water, try a candle.

Which melts the fastest?

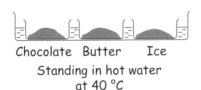
Chocolate Butter Ice
Standing in hot water at 40 °C

Cheese

Safety!
Discuss the hazards and risks associated with this activity. The children doing this activity, should be well supervised.

Limit the children to a few materials that will melt at the temperature of very hot water (80 °C) or below, such as soft and hard fats, chocolate and wax. Use small, equal cubes of each substance and place each one into a dish. Place these into a large container of water with a thermometer. Begin the activity with warm water (30 °C) and see if any substance begins to melt. Remove the dishes, add hot water to the water bath, stir and take the water temperature. Replace the dishes to see which substance melts at the higher temperature. Record the water temperatures as different materials melt. A bar graph can be made of the results.

HEATING AND COOLING MATERIALS

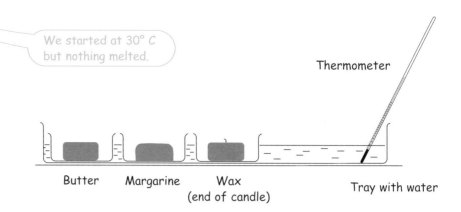

We started at 30° C but nothing melted.

Graphing package

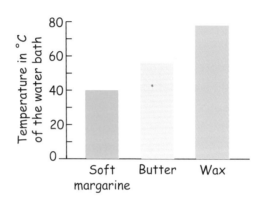

A graph to show the temperature at which materials begin to melt

Ice Balloon

How long will it take to melt?

I can hear the ice cracking.

(d) **More melting ice** Great fun for younger children. Fill different containers with water eg balloons, rubber gloves, ice trays, yoghurt pots etc, freeze them and present them to the children to watch them melt during the day. You can peel the balloon away from the ice to leave the smooth, ice ball. Ask them how the process might be speeded up or slowed down (thermal insulation), and which they think will melt first. Float the balloons on warm water. You could use the concept cartoon *Snowman* here (see page 32).

(e) Older children might **investigate** 'Does salt affect melting ice?' Discuss the use of salt and grit on icy roads then carry out an **investigation** which they may plan themselves, or direct the activity. Prepare two beakers and funnels, put the same quantity of ice in each funnel. Pour salt on one lot of ice and time the melting as the water drips into the beakers. (For more able pupils a thermometer can be put into the ice cubes to note the temperature change). As an **investigative activity,** have two dishes each with the same quantity of ice cubes on. Pour salt on the ice the top of one dish and compare the melting of the cubes with the dish without the salt. Repeat the activity or do it at the same time and have a dish with the salt underneath the ice cubes rather than on top of them. Get the children to observe closely where on the cubes the melting begins.

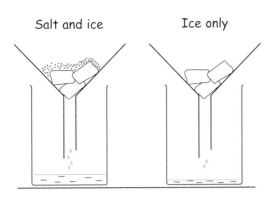
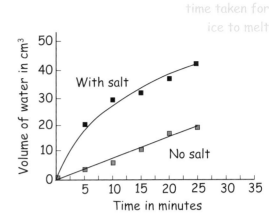

A graph to show time taken for ice to melt

(f) **Cooking** A perfect opportunity to observe, describe and with older children explain and investigate the changes that take place during cooking activities (see recipe sheets, page 86).

- Pastry cooking, eg jam tarts.

- Bread baking, investigate the effect of warm or cold water on mixing the yeast.

- Eggs cooked for different lengths of time or in different ways.

- Drop egg white using a dropper into different temperatures of water to see at what temperature it changes. Drop it into cold water first then into very hot water, try to find the critical temperature. An investigation for older children, 'At what temperature does egg white change?' (Because boiling point changes with pressure, it is not possible to cook an egg on the top of Mount Everest because the white will not change.)

- Cake making, make little cakes and investigate 'what happens if you leave them in the oven for different lengths of time?'

- Discuss the reversibility of these changes with older children. Are any new materials made?

(g) **Clay** As part of an art activity, children can explore clay as a material and model with it. As well as firing the clay, some can be left to dry out so that a comparison can be made.

(h) **Wax** Wax can be used for a variety of art activities. Old wax, candles or crayon stubs can be melted in a container over hot water. (Grating it first speeds up the process.) This can then be poured into new moulds and while it is solid but still soft and pliable it can be re-worked. Using an old paintbrush, the molten wax can be painted onto paper (or drip wax from a candle) then colour washed with inks or paint. The hard, dry wax can be picked or ironed off.

Wax may be painted onto fabric then dyed and when all is dry the wax is ironed out as in the 'batik' technique, page 19.

HEATING AND COOLING MATERIALS

Safety!
- Spirit thermometers should be used.
- Care with handling glassware.
- Care when handling hot and cold things.
- Candles should be stood in sand or water filled metal trays.

- Drip wax over paper, leave to dry.
- Wash with brush and ink or paint.
- Pick off dry wax.

Wax drip pictures

i **Wax candles** Different sized and coloured candles can be made using the stubs of old candles or specially bought wax. Wicks and wax colour tablets can also be bought from craft shops (see resource list). Used washing-up liquid containers, cut down, make good candle moulds.

ii **Dipping candles** Make up a molten coloured wax mix using plain wax and a colour tablet. Using a plain, white household candle and masking tape, create a pattern on the candle, then dip the candle into the coloured wax.

iii **Ice candles!** As a demonstration or with well supervised children, but great fun! For a mould use a cut down washing–up liquid container and have ready an old candle cut to the same height. Melt some wax (old candles are fine) and pour a little into the bottom of the mould to make a base. Stand the candle in this and hold until firm. Pack the space between the candle and mould with broken but not crushed ice cubes. Now pour the rest of the molten wax over the ice to fill the mould. Leave it to stand for half an hour. When it is cool and firm, run under a warm tap to loosen the wax and push firmly out of the mould. You have a candle like a Gruyere cheese! It is worth making one to show the children first and asking them how they think it is made. A lot of discussion about changes of state of different materials is possible here.

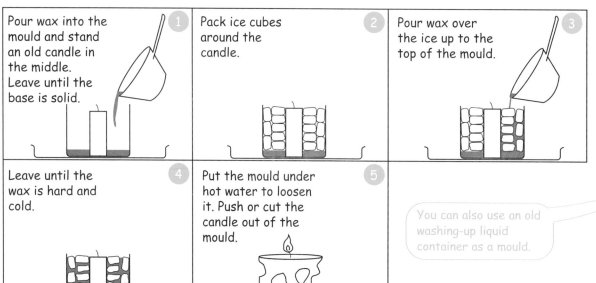

1. Pour wax into the mould and stand an old candle in the middle. Leave until the base is solid.
2. Pack ice cubes around the candle.
3. Pour wax over the ice up to the top of the mould.
4. Leave until the wax is hard and cold.
5. Put the mould under hot water to loosen it. Push or cut the candle out of the mould.

You can also use an old washing-up liquid container as a mould.

(i) **Expanding gases** You need a bowl of hot water, a bottle and a balloon that will fit over the neck of the bottle. Discuss with the children what is in the bottle, put the balloon over the neck and stand it in the hot water, the balloon inflates. Then take the bottle out and watch the balloon deflate. Investigate... 'Does the size of the bottle affect the amount the balloon inflates?' Or 'Does the temperature of the water affect the amount the balloon inflates?'

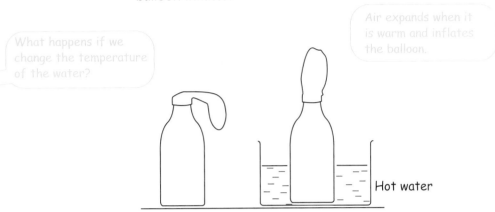

(j) **Expanding Liquids** Fill a small bottle up to the top with coloured water, insert a clear straw and secure in place with plasticene to make a seal. Stand the bottle in hot water and watch the liquid rise up in the straw. If you use a plastic bottle the water goes down first then rises, due to the initial expansion of the plastic. (Glass does not expand very much.)

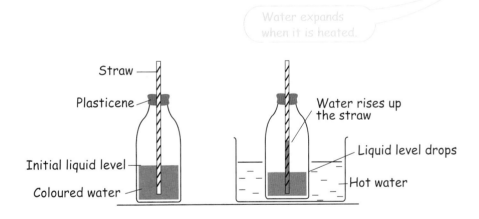

(k) **Making a thermometer** Ideally the children would do this activity after using a commercial thermometer and doing the above activity on expansion of liquids. Half fill a small bottle with cold, coloured water. Mark a clear, plastic straw with 1 cm intervals and place it in the neck so that it touches the liquid and secure in place with plasticene. Push down on the plasticene so that water begins to fill the straw. As the water in the bottle warms up the water rises up the straw like a thermometer. This can be dramatically shown, by standing the bottle in hot water. Remember, with plastic bottles there is initially a drop in the water line because the plastic expands.

(l) **Liquids evaporate when we heat them** Heat water in a heatproof container gently over a candle and watch it evaporate as the liquid turns to a gaseous state. If you have Pyrex® beakers the liquid level can be marked at timed intervals. Repeat and compare with other liquids such as lemon juice. (Vinegar gives off a very strong smell when heated.)

(m) **Water boils at 100 °C to produce an invisible gas called steam** A demonstration for older children is to watch water boiling in a saucepan on a cooker or in a kettle. The teacher should take the temperature of the water as it boils. (Use a thermometer that measures to 110 °C.) Discuss the bubbles and the 'steam' they see. If you also use a kettle, there is a gap (observed against a dark background), due to the invisible steam, immediately above the kettle spout. Take the temperature here then higher above in the water vapour cloud. As a safety issue discuss the hazards and risks with the children.

Some materials change on cooling

(a) **Change of state-liquid to solid** This activity is the converse of activity (b) 'More melting ice' in the previous section and could precede it. Get the children to fill different shapes with water and put them in the freezer overnight. They should also explore freezing other liquids such as milk, oil, juice and vinegar.

(b) **Water is an unusual substance, it expands when frozen** Older children could investigate the change in volume of different liquids including water as they cool down from very hot to frozen. As most liquids that they are likely to come into contact with contain a large proportion of water, they will find that many will expand on freezing. This is a good discussion point. This is also a good opportunity to construct a line graph of temperature and volume. Using a glass container initially will also minimise the effect of the hot liquid on the material of the container. Half fill a small, glass bottle or jar with hot water, cover it to prevent loss through evaporation and mark the liquid line. As it cools, note the temperature and any change in the liquid line. To go below room temperature, the container can be put in the refrigerator or stood in ice. Finally, put it in a freezer.

Narrow, glass jars are the best for showing volume change.

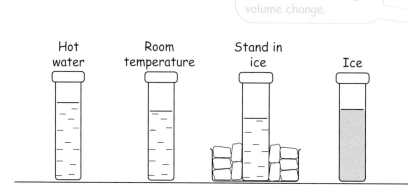

(c) **Ice damage** This activity could lead to a good discussion about the damage done when water seeps into rocks and bricks, then freezes and expands causing them to crack. This also happens to burst pipes. Fill a plastic bottle to the top with water, put the top on and freeze it. Some plastic bottles are more flexible than others and will 'give' more, try to use a rigid one and it will split. Try putting a small screw top glass jar full of water in a clear, plastic bag and freezing it.

(d) **Steam and water vapour cool to produce water, this is called condensation** This demonstration may be done at the same time as the above activity on 'Water boils at 100 °C to produce steam' in the previous section. Put a large mirror or cold shiny saucepan above the boiling water and watch the steam turn to water and the condensation appear on the shiny surface. This can also be done with water that the children gently heat over a candle themselves. It does not have to be boiling water to evaporate and condense on a mirror. It works best if the mirrors are refrigerated for a while first. Traditionally, we get children to breathe on a cold mirror, the problem with this is the initial understanding that our breath contains water. This therefore may be done after the previous activity with the question 'what does our breath contain?'

(e) Put crushed ice in a glass jar and cover the top to show that water is not escaping! Leave the jar for about 10 minutes (depending on the outside temperature). As the warm water vapour in the air touches the cold glass, it condenses leaving the outside of the glass wet. Note that some children are still convinced that water from the ice is the cause of this! Also try cold cola cans from the refrigerator, an everyday experience for many children.

(f) Put a plate of ice over a transparent bowl of warm water and watch the warm water vapour hit the cold plate, condense and drip back into the bowl.

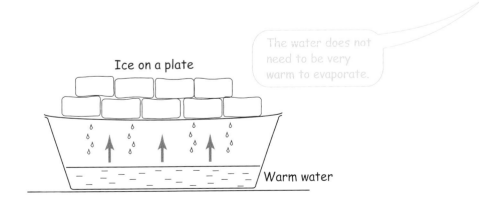

Ice on a plate

The water does not need to be very warm to evaporate.

Warm water

HEATING AND COOLING MATERIALS

star Poetry
by Michael Rosen

Chippy Breath Writing in the condensation on bus windows.

Hot Cat Different things melt at different temperatures!

Some Thoughts About Eggs Eggs can be cooked in a variety of ways.

Grandad's Airer Clothes drying on an airer.

Hot Pants Hot air dries clothes in a tumble drier.

Steamy Shower Evaporation and condensation in the bathroom.

Thirsty Land The sun dries up the water on the land.

Scoop a Gloop Materials change when they are heated, clay is a permanent change. See page 9.

Chippy Breath

After football
my dad buys me fish and chips
and my hot chippy breath
makes clouds in the air
and rain on the windows
of the bus
all the way home.

I write the score
on the wet glass
– but only when we win.

Hot Cat

The butter melted
the cheese went mouldy
it all got so hot
the cat moulted.

Then it got hotter
the cheese melted
and the cat went mouldy.

Then it got hotter
and the cat melted.

Some Thoughts about Eggs

1. Is a hard-boiled egg one that was hard to boil?
2. How fast do runny eggs run?
3. It can't be very fast because my dad beat one.
4. If the white bit's called the 'white', why isn't the yellow bit called the 'yellow'?
5. If a piglet is a little pig, is an omelette a little om?
6. What is an om?
7. You put mushrooms in mushroom omelettes, you put cheese in cheese omelettes. I'm worried about what they put in Spanish omelettes.
8. Why don't eggs melt when you heat them up?
9. Who lays Easter eggs?
10. If we had an egg we could have egg on toast – if we had some toast

HEATING AND COOLING MATERIALS

Grandad's Airer

My Grandad says
when he was a boy
he lived in a flat
and in winter
when they wanted to dry the washing
they used an 'airer'.

It was like a clothes horse
lying on its side
that they hoisted up to the ceiling
with a rope and pulley.

The trouble was, he said,
it was next to the cooker.

Just where
they fried liver and onions.

So when he stood in line at school
the kids behind would sniff his shirt
and point and say:
'Liver and onions!'

Steamy Shower

I love a
dreamy, steamy shower
hanging about
for over an hour
just before bed
getting hot and red
in the steam
standing there
with time to dream
water-running-over-me feeling
drips dripping off the ceiling
mum says it's my fault it's peeling
nothing can beat
the hot wet heat
nothing wetter
nothing better
I love a
dreamy steamy streamy shower

Hot Pants

The tumble drier
dries socks hot
and hot socks
make my toes warm.

All through winter
when it's wet and cold
our tumble drier
rumbles round.

Hot socks
hot shirts
hot skirts
hot pants

All through winter
in the wet and cold
I watch where the pipe
from the drier ends:

It's where there's a grille,
and through the holes
the drier breathes out
hot air.

Hot air
hot breaths
hot puffs
hot pants.

Thirsty Land

In the plane over the desert
I see the badlands beneath us
wrinkled like old skin.

Ancient river valleys
are now dry brown snakes.

The sun that glares at this
has dried up
every last drop.

Even the rocks begin to say:
'Water, please, water.'

COOKING RECIPES

Biscuits

125 g butter/margarine
150 g caster sugar
1 egg yolk
225 g plain flour
grated lemon
mixed spices or vanilla

Cream the fat and sugar until white, then add the egg yolk and beat. Stir in the flour and flavouring. Knead lightly, roll out and cut with a cutter. Put on a greased baking sheet and bake for 15 min at 180 °C (350 °F, gas mark 4). Take a few biscuits out at 5 minute intervals to see the effect of time or bake at different temperatures.

Bread

700 g strong plain flour
2 teaspoons of salt
1 teaspoon of fat or oil
1 teaspoon dried yeast
1 teaspoon caster sugar
400 ml tepid water

Mix about 100 ml water with the sugar, add the yeast and whisk. Leave until frothy. Mix the flour, salt and fat then add the yeast mix. Knead until smooth then leave in a covered bowl to rise in a warm place. Knead the dough again and put into a greased loaf tin or divide into small pieces for rolls.

Investigate water temperatures for the yeast and make a batch of bread without it.

Cakes

100 g butter or margarine
100 g caster sugar
2 beaten eggs
100 g self-raising flour

Cream the fat and sugar until white add the eggs and beat. Fold in the sieved flour. Put into small tins or cases and bake for 15-20 min. Investigate different cooking times on some of the cakes or investigate using baking powder as a raising agent instead of S.R. flour. (225 g plain flour : different amounts of baking powder.)

BURNING

Science background for teachers

VOCABULARY

Heating, burning, reversible, irreversible, oxygen, carbon dioxide, chemical reaction, fuel, hazard, toxic, temperature, flammable, ignite

Children often confuse the terms 'heating' and 'burning' so it is worthwhile clarifying the difference with them. When something burns a flame is usually seen but this is not the case with heating. Burning is a chemical reaction, a chemical change. Heating is raising the temperature of a material and this can be reversed. Burning is a chemical process, is not reversible and a new material is formed as a result. For example common fuels such as wax produce carbon dioxide, water and soot. For burning to take place there has to be fuel, oxygen and a high enough temperature, this is often called the 'fire triangle'. At normal room temperatures most materials will not burn because the fire triangle is incomplete, it takes heating to raise the temperature of the fuel enough to start the fire. Forest fires start because of the intense heat of the sun which raises the temperature of easily combustible material such as very dry grass (as the fuel) to make the conditions right. When we strike a match this causes friction and some of the chemical energy is transferred to heat energy and the combustible material in the head of the match ignites. When this is done in the presence of a fuel, burning takes place. Different materials require different temperatures to ignite which is why some burn more easily than others do. Some are so unstable that they will ignite at fairly low temperatures. A few materials such as white phosphorus ignite at room temperature. It is stored under water to keep out the oxygen and as soon as it is removed from the water it bursts into flames. Other materials, for example most metals, require extremely high temperatures to burn.

A candle is a good example to observe and consider when understanding the burning process. Both the wick initially and the wax are the fuel. The burning wick supplies the heat to the main fuel which is the wax, which melts around the wick, vaporises then burns at the top of the wick. Capillary action then causes more molten wax to rise up the wick and the flame begins to glow brighter as a result of the increased fuel supply. (This can easily be seen in a glass spirit burner.) If the wick is too small it will produce insufficient heat to ignite the fuel and the candle goes out. A wick that is too big produces too much wax vapour for the available oxygen to burn it, the burning process is incomplete and a sooty smoke is produced. Children often confuse gas and smoke, both of which may be produced as a result of burning, but smoke is a mixture of solid particles suspended in a gas, eg carbon suspended in air.

The natural gas that is burnt in our homes is methane.

In addition to fuel and heat, oxygen is required. This is demonstrated when a fire is extinguished by smothering it to block off the oxygen. A burning candle will eventually go out if you cover it with a beaker because the oxygen supply has been cut off. If the burning candle is stood in a dish of water, the water rises up inside the beaker until the candle goes out. Part of the explanation for this is that the water replaces the oxygen that has been used, but it is not that simple and by no means the whole explanation. Initially when the beaker is put over the candle, the warm air expands and may bubble out into the water. Carbon dioxide and water vapour are produced as a result of combustion and oxygen is used up. Some carbon dioxide dissolves in the water and the water vapour condenses on the inside of the beaker occupying much less volume than before. The pressure inside is then less than outside, and water is forced into the beaker.

It is important to discuss the hazards of burning with children so that they understand the dangers. Older children can discuss the hazards and risks of some of the activities. Some materials burn very easily and others, especially some plastics, produce toxic fumes, so children can look at furniture and fabric hazard labels. Children should be made aware of situations in their lives that are based on the scientific facts that they learn; for example, opening a window in the presence of a fire increases the oxygen supply and fans the fire, smothering it with a blanket can put it out. Water on a fire lowers the temperature and will put it out, especially if it is a solid fuel fire. However there are dangers with water and certain types of fire and children need to understand the reasons for this. Water is a conductor of electricity and if sprayed onto a fire caused by an electrical fault may cause an electric shock. Oils and petrol do not mix with water and may float on the surface. If they are burning, they will continue to do so and spread further. Hot fat has a very high temperature and will turn water into steam immediately and explode, sending the fat flying everywhere, where it could easily ignite.

SKILLS
- Using equipment carefully and working cooperatively.
- Understanding, planning and carrying out a fair test.
- Observing and recording with accuracy.
- Constructing a line graph.

BURNING

Key ideas and activities

When materials are burned a new material is formed and the process is not reversible

Safety!

Discuss the hazards and risks with the children before doing any of the activities to highlight potential dangers.
- Carry out activities in small groups or use extra classroom assistance.
- Be aware of the location of fire extinguishers or blankets. A bowl of water is a useful standby.
- Goggles or safety glasses should be worn, long hair tied back.
- Care when burning any material, the activity should be closely supervised.
- Hot plastics are dangerous because they melt and are sticky when heated or burnt. Many plastic materials give off seriously toxic fumes, and should be avoided.

(a) **Prepare a selection of different materials for the children to burn and observe** (This could be carried out as a demonstration.) Only very small quantities (a 2 cm square) are needed. The children can either hold them in metal tongs or set them alight in small metal containers standing in a sand tray. Individual foil baking cases, old baking trays, or the metal containers from used nightlights are suitable for this. Some suggestions for materials are: different papers, dry twigs or dead matches, sawdust, straw, cotton material, polyester, wool, birthday candle, wire wool, steel nail, flame retardant material. The children need to closely observe what happens and chart the results taking into account what is left at the end.
Were there any smells given off?
How easily did it ignite?
For how long did it burn?

Be safe! Wear goggles.

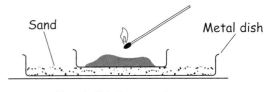

Metal dish in a sand tray

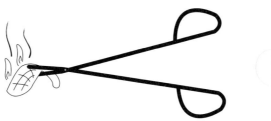

I can see smoke and it smells.

(b) **Investigate** whether the weight of a material changes after burning, is some of the original material 'lost', if so where? This should be closely supervised and is for older, more able children. Repeat the above activity using only one or a few materials to investigate. Weigh the material before and after burning using electronic scales accurate to at least one decimal place. Avoid synthetics for this activity as a larger quantity of material is needed to register a reasonable weight and they give off toxic fumes.

Material	Did it ignite easily?	How did it change? Observations	Any smells?	Burning time in secs	What is left?
Paper					
Cotton fabric					
Wire wool					
Wood spill					
Tin-foil					

BURNING

(c) **Burning a candle** Use a household candle or a nightlight and get the children to closely observe it burning at all stages and note down what is happening. Discuss this with the children to explain the molten wax being absorbed by the wick. Showing them a small, glass spirit lamp might help this concept.

> The wick burns first and melts the wax.

> The melted wax is absorbed by the wick. It vapourises and then burns.

Safety!
- Candles should be stood in a metal tray in sand or water.

> What has happened to the weight of the candle?

Air (oxygen) is needed for burning to take place

(a) Repeat the activity **Burning a candle**. Allow it to burn for a minute or so and then place a heatproof beaker or a big glass jar over the top (so the flame doesn't touch the glass). Carefully observe what happens. Why does the candle go out? What do you see on the inside of the jar? This could be an investigation that the children plan themselves. Investigate that burning needs air (oxygen) to take place.

Water droplets

> Why are there water droplets on the inside of the beaker?

> Why does the candle go out? Will it burn for longer in a bigger beaker?

(b) **Investigate** 'Does a candle burn longer in a bigger beaker?' The children can plan this investigation for themselves and construct a bar graph of the time the candle burns and the size of the different beakers.

> Use the same type of candle each time. Nightlights are very safe and work well.

Graphing package

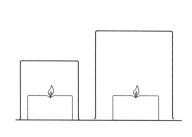

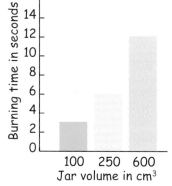

Burning time in seconds vs Jar volume in cm^3 (100, 250, 600)

(c) **Investigate** 'Is twice as much air (oxygen) needed for two candles to burn, three times as much for three candles to burn?' This requires careful timing of the candles burning in the same size beaker (jar), so choose a big one to start with or use birthday candles stuck into plasticene. The more candles in the jar, the quicker they go out, because the more oxygen they use.

(d) **Candle, beaker, water activity** This could be a directed activity or a demonstration. Stand the candle in a dish of water and allow it to burn for a minute. Then cover with a beaker or jar and observe the water enter the beaker. **Investigate** 'Does the water increase in proportion to the number of candles burnt?'

Try burning more than one candle.

Birthday candles in plasticene

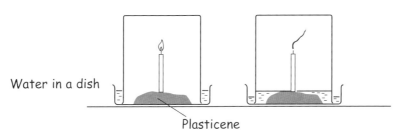

Water in a dish

Plasticene

To identify and assess the risks associated with burning and understand the consequences of actions taken to control burning

(a) Look at different materials and everyday situations and discuss safety in relation to this. Children could look at different 'home situation' pictures and discuss the burning hazards there. Look at hazard labels from fabrics to recognise the potential dangers, for example standing near a fire wearing a nightdress.

(b) From their own experience in their 'burning activities' and from class discussion, children can list materials that burn easily, and which give off toxic gases when burnt. They could design a poster or leaflet illustrating these dangers for a specific audience eg 'The under fives'.

(c) Look at the labels and contents of different types of fire extinguishers to raise their awareness of treating different types of fires in different ways. Discuss this with them.

(d) Children could design a safety poster showing the ways in which different fires may be treated.

by Michael Rosen

Barbecues Barbecues are fine in your own garden, other peoples pollute the air!

Berlam Bam Boola Burning driftwood changes shape and only ashes are left behind.

Barbecues

When it began to drizzle
I thought the whole thing would fizzle
out.
But there was no need to grizzle
about it, because soon the stuff started to sizzle
and in the end it was
GREAT!

I love it when **we** have barbecues
in the garden but it really doesn't amuse
me when the next door neighbours use
theirs. I refuse
to believe our one makes as much
smell and smoke as their one –
which **I HATE!**

Berlam Bam Boola

That night
we made a fire
on the beach
and Alan danced about
singing:
Berlam bam boola
Berlam bam boola
tooty fruity

I found
a bit of drift wood
that looked like
a cow's skull
and it burned up bright
Berlam bam boola
berlam bam boola
tooty fruity

When we found
the remains of the fire,
in the ashes
I could just make out
the shape of a cow's skull.

Links to History and ideas & evidence in science

Victorian Britain – Humphry Davy (1778-1829) and the safety lamp

Sir Humphry Davy was an English chemist and much of his important work in the first decade of the 19th century focused on the relationship between chemistry and electricity.

At this time, mining was an important industry, providing coal to power the new machines and for domestic heating. It was however a dangerous occupation due to the frequent explosions which occurred underground. The decomposition of plants to form coal, which takes place over millions of years, produces a gas called methane. It is trapped in the coal strata and escapes when the coal is mined. Methane mixed with air is known as 'firedamp' by the miners. It accumulates in the mine galleries and is explosive. Before the electric light, the only means of illumination underground was candlelight or oil lamps, which would cause the 'firedamp' to explode. Men were often thrown up through the mineshaft, pit props collapsed and there was a tremendous loss of life.

In the August of 1815, after a particularly serious accident, Davy was approached by a safety society to investigate the problem. He discovered that methane was the least readily combustible of all the inflammable gases and required the highest temperature to ignite. It did not explode when in contact with red hot iron or charcoal and the heat produced from it when it burnt was less than that from other inflammable gases. He found that if the gas was ignited in a narrow glass tube or particularly a metallic fine mesh tube, it was contained and did not explode. The metal wires quickly conduct heat away from the hot gases passing through the spaces of the mesh and it becomes too cool to burn on the other side of the mesh.

He designed a safety lamp, which was made of copper gauze and burnt Greenland whale oil. In the presence of methane, it does not explode but the appearance of the flame changes, thus also acting as a 'gas detector'. This was used in the pits in the January of 1816 with great success. Explosions occurred then, only because the miner's removed the gauze! This was because the illumination wasn't that good. So the lamp was eventually refined in design and although electric lamps are used today, the safety lamp is still used as a 'gas detector'.

This topic of burning can be linked to other curriculum areas, as shown in this example.

BURNING

The Original Safety Lamp design

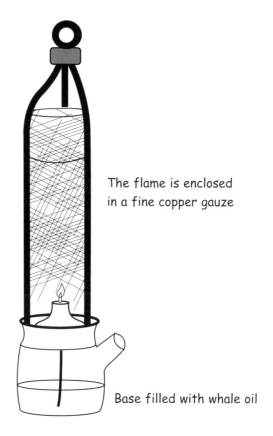

The flame is enclosed in a fine copper gauze

Base filled with whale oil

IRREVERSIBLE CHANGES – chemical reactions

Science background for teachers

VOCABULARY

Reaction, reversible, permanent, irreversible, chemical, iron oxide, rust, corrosion, vinegar, plastic, milk, carbon dioxide, borax, Plaster of Paris

Most changes that cannot be reversed are chemical reactions where a new material is formed, and it is not possible or extremely difficult to recover the original materials. Children will experience such changes all the time in their everyday life and in the science activities they do in school, but it is not always obvious that a chemical reaction has taken place. Changes that take place in cooking, some heating, mixing some materials, such as vinegar and bicarbonate of soda, and burning are all chemical reactions. As children experience these activities it is worth discussing this with them as an on-going idea so that they begin to develop and build on the concept.

In a chemical change, the bonds between the particles of the substances are broken and reform in a different arrangement as a new substance. This may require energy taken from the surrounding material, and there is a drop in temperature as happens when lemon juice and bicarbonate of soda are mixed. Very often however, there is energy produced as a result of the reaction and an increase in temperature, as in the reaction between Plaster of Paris and water.

Rusting and the burning of a fuel in the presence of oxygen are both chemical reactions known as oxidation. Iron, in the presence of water and oxygen, reacts to form iron oxide or rust. Whilst the rust can be rubbed off to reveal the 'good' iron underneath, the top layer has been changed and removed permanently. Since this change is one that children will see happening and causes such a lot of corrosive damage, it is worth investigating the factors that affect it and what may prevent it from happening. Steel rusts because of its iron content. Stainless steel does not because traces of other metals have been added, this prevent the oxidising process which gives rise to the rusting. Galvanised steel or iron is coated in zinc, which also prevents rusting. Old copper coins react with the air and copper oxide forms making them 'dirty' and black. If they are dropped into cola, the strong phosphoric acid in the cola reacts with the copper oxide and cleans the coins.

Other chemical changes Mixing materials together often causes a chemical reaction and it is interesting for children to do some examples of these and compare them with mixtures that only change physically. Salt dissolves in water to form a solution, a physical change that can be reversed by causing the water to evaporate leaving salt crystals behind. Alka-Seltzer also dissolves in water to form a solution, but there is also a chemical change and carbon dioxide, a new material is produced. This is a chemical change that cannot be reversed.

IRREVERSIBLE CHANGES

SKILLS
- Ability to plan and carry out a fair test.
- Use equipment accurately, with care and work cooperatively.
- Observe carefully and record results.

Key ideas and activities

IRREVERSIBLE CHANGES

Non-reversible changes result in the formation of a new material that may be useful

Making plastic with milk and vinegar This can be a directed activity or a demonstration. Warm about half a pint of whole milk and add about 1 tablespoon of white vinegar to it. The milk immediately curdles. Strain and keep the solid curds and put the plastic on kitchen paper to dry it. You can then handle it, put it in a mould or shape it yourself. This will dry firm overnight if kept in a warm place, become very hard in time and can be painted and varnished. Milk can be kept warm in a thermos flask. The temperature and exact quantities are not critical for this activity.

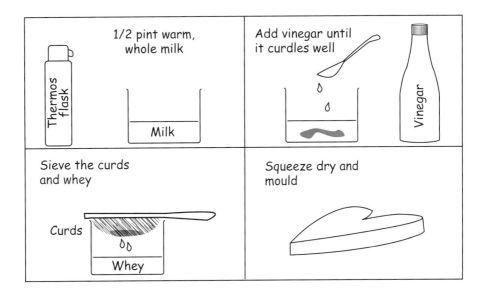

Making plastic with PVA glue and Borax This will make a similar but translucent plastic. (Borax can be bought very cheaply from a chemist.) Make up a small quantity of saturated borax solution (this is when no more solid will dissolve in the water, there maybe a little borax left in the bottom of the container). Add a teaspoon at a time of this solution to one tablespoon of PVA glue. Mix continuously and the glue will absorb the borax solution. Keep adding and mixing as the PVA changes to a solid that is 'bouncy' and can be handled and moulded. Leave to dry for a few days in the warm, it will go very hard. This is a good group activity, as a small jar of saturated borax solution can be used by a group of six children each with their own spoonful of glue in a container.

Mix the borax solution really well into the PVA glue. Add more until it changes.

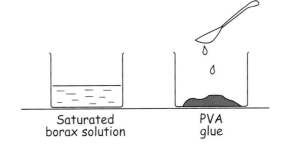

IRREVERSIBLE CHANGES

- Care when handling glassware.
- Care with iron filings, they are a potential hazard for eyes.
- Care with sharp objects eg nails.

Making a model with Plaster of Paris and water When water is added to Plaster of Paris there is an exothermic reaction, heat is given off as the plaster is setting and gets really warm as the plaster gets hard. This may take 15 minutes and can be clearly felt or measured by older children with a thermometer, but do not leave the thermometer in it too long or it will get stuck! Make a mould with a card strip, paper clip and rolled out plasticene. Make an impression on the plasticene and grease it lightly with Vaseline. Pour in the mix of water and Plaster of Paris, which should be the consistency of thick cream. Leave this to set.

Making carbon dioxide There are various ways of making this, (refer also to Solids, liquids and gases section). Perhaps the most spectacular is with sodium hydrogencarbonate (bicarbonate) or baking powder and white vinegar, to which you can add food colouring or paint. When the vinegar is added to the baking powder, the reaction is immediate, carbon dioxide is produced and makes a lot of froth. An active volcano can be made by making a mountain using damp sand, and gravel in a tray. The dry powder is put into a small jar or plastic bottle in the centre of the mountain, then the vinegar and red food colouring mixture is dropped onto it. It flows out of the jar as lava!

For a quick, explosive reaction that will blow the lid off a film canister, try half an Alka-Seltzer tablet and water. (See page 58.)

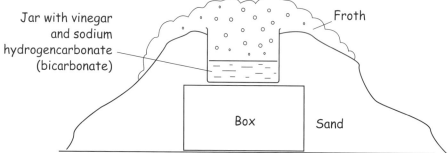

A 'sodium bicarbonate' volcano

Some irreversible changes are not useful and may be corrosive

Rusting Discuss with the children the general idea of rusting and show them something that has rusted. Discuss where they see it, what they think causes it and the materials that rust.

Referring back to the rusted object you showed eg an iron nail, this material can then be used as a basis to begin the activities on the factors that affect rusting and its prevention. The material they have seen with rust on is then eliminated as a variable. They know that iron rusts so they can go on to find out what conditions make it happen. The rusting process may take several days. If iron filings are used it only takes about an hour because there is a larger surface area. Children could use iron filings and nails to do a comparison of the time it takes.

(a) **Investigate what makes things go rusty?** This can be a simple, directed activity or a planned investigation. If the children are planning the investigation, they need to begin by deciding what factors they think make things go rusty, and setting out to prove it. As a directed activity, there are a variety of ways to carry out this activity but it can simply be by putting a nail in a dish on damp cotton wool and another on dry cotton wool. This shows that rusting takes place in the presence of water and air. Eliminating the air/oxygen completely is difficult. Boiling the water for 5 minutes to force out the air, putting this into a jar with the nail and a layer of oil on the top of the water is often successful. However, a little rust sometimes appears and it is quite difficult to explain this to the children.

(b) **An activity to show that rusting uses up oxygen** This directed activity or demonstration can be used to show that air/oxygen is used up in rusting, and works well. Wet the inside of a test tube and sprinkle in some iron filings, which then stick to the wet tube. Invert the tube into a beaker of water at an angle. As the iron filings rust, oxygen in the tube is used up and water rises up the tube to replace the used oxygen. It begins to work after about an hour and works really well after a day or so. If the tube is marked into five equal sections before beginning the activity it can also be used to show the approximate percentage of oxygen in the air. The water rises about 1/5 of the way up the tube.

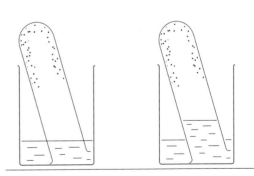

Water rises up inside the tube as the iron filings rust.

(c) **Investigate does the type of liquid affect rusting?** As a directed activity, the children can prepare cotton wool soaked in variety of liquids eg distilled water, salty water, sugar water, vinegar, oil and tap water. Each piece of cotton wool is then placed in a dish with an iron nail and they wait to see how much rust forms and how quickly over a period of a week. As an investigation, the children plan the liquids they want to try, and the way they want to do the investigation.

(c) **Which materials rust?** This is a simple, classifying activity which can be directed by the teacher or planned as an investigation by the children. As an investigation the children need to decide which materials they are going to test, and how they are going to get them to rust. They will have some ideas about rusting from investigating the factors that affect rusting, this will help them to plan how to get the materials to rust. If it is a teacher-directed activity give the children a selection of small objects made of a variety of materials eg safety pins, paperclips, brass screws, coins, iron filings, steel wool, plastic object, rubber, wood etc. Put these materials in conditions that would induce rusting such as on damp cotton wool in a dish.

Word processing

Note that in warm conditions evaporation will occur and the cotton wool may need 'topping up'.

Object to test	Material	Results after...				
		1 day	2 days	3 days	4 days	5 days
eraser	rubber					
screw	brass					
paperclip	steel					
nail	iron					
pentop	plastic					

(d) **Investigate ways of preventing rusting** Discuss the issues of protecting iron that has to be exposed to air and water and discuss ways in which it might be protected. Get the children to plan their own investigation using any of the ways previously used to rust materials and get them to decide on ways to protect the materials. Any oil or grease based material will help to prevent rusting, they could see which was the best eg Vaseline, model paint, engine grease, cooking oil.

by Michael Rosen

Concrete Paw Marks left in concrete are permanent.

Rust Some advice for you if ever you are thinking of walking across a very old bridge.

Some Thoughts about Eggs Cooking eggs in different ways. See page 84.

Scoop a gloop You can change the shape of clay, heating it changes it permanently. See page 9.

Drizzy Fink Confusion about fizzy drinks! See page 60.

Concrete Paw

Our friends next door
have moved away.
They put everything
in a van today.

They got in a car
and went.
That's it.
No one's left.

My friend's gone.
Her sister's gone.
Their mum's gone.
Their dog Sniffy's gone.

Everyone's gone.
Nothing left...

...except for
the mark of Sniffy's paw
printed into the concrete
by their front door
where it'll stay
forever more.

Some advice for you if you are thinking of walking across a very old bridge

Don't trust
rust

this page has been intentionally left blank for your notes

THE WATER CYCLE
evaporation and condensation

Science background for teachers

Water is one of the most important substances on Earth, being essential to all living cells, 75% of our bodies is made up of it! Water covers 70% of the earth's surface and is the most common liquid we come across, 97% is salty and 3% is fresh. Of the fresh water, 85% is in the form of ice and the air we breathe includes water vapour, the gaseous state of water. It is continually recycled on our planet. Water from the earth's surface in the form of seas, lakes, rivers, down to puddles evaporates and returns to the atmosphere as water vapour. This becomes cooler, condenses to form tiny water droplets or ice crystals and forms clouds. When these water droplets become too heavy to remain floating in the air, they fall as rain or snow and this is known as precipitation.

Evaporation

VOCABULARY

Liquid, gas, gaseous, water vapour, steam, evaporation, condensation, precipitation, water cycle, conditions, rain forest, ice, crystals, transpiration, root system, cells, specialised, humid, atmosphere, floating, air stream, molecules, perimeter, surface, treatment, waste, sewage, reservoirs

When water boils at 100 °C it changes from a liquid into a gas that we call 'steam' and moves into the surrounding air. However, boiling does not need to occur for evaporation to take place. In any quantity of liquid, the molecules at the surface have less interaction with each other than those in the body of the liquid. These molecules leave the surface and evaporate into the air. This gas is not hot like steam, but cool and is called a vapour. As it happens at the surface and is dependent on temperature, the larger the surface area and the warmer it is the faster it happens. If a wind or stream of air is present to blow away the slowly evaporating liquid molecules, they move away faster allowing the next layer of liquid to evaporate, so speeding up the process. So large, shallow, puddles dry up faster than narrow, deep ones containing the same volume of water. They all dry up faster on a warm, sunny or windy day.

Another effect of evaporation is that of cooling. As surface liquid evaporates, energy is transferred from the liquid to the vapour resulting in the liquid becoming cooler. This happens in sweating, as the sweat on the surface of the skin evaporates the skin cools down. The faster this happens, the more dramatic the effect, eg surgical spirit on the skin evaporates very quickly leaving the skin feeling very cold.

Condensation precipitation and the water cycle

As water evaporates from the oceans, seas and rivers, which are two thirds of the Earth's surface, the air becomes filled with water vapour and may become humid. If this continues, it becomes saturated with water vapour and can hold no more, but this depends on the air temperature. The higher the temperature, the more water vapour can be held. As saturated air cools down, the water vapour changes back into the liquid state or condenses and then forms tiny water droplets in a cloud, mist or fog. If this is near the ground it is a mist or fog, but clouds form well above the ground and if they are cold enough, they consist of ice crystals. As these ice crystals or water droplets bump together and get bigger, they become too heavy to remain floating in the air on the upward air stream and they fall. This is called precipitation. If the air is warm, the ice crystals melt and fall as rain, if not, they fall as snow. This process of evaporation and condensation is the water cycle.

Transpiration and the water cycle

Plants also contribute to the water cycle during a process called transpiration. Water is taken up by their root system, is transported in specialised cells in the stem, travels up to the shoots and leaves where it evaporates. Trees lose vast amounts of water through transpiration, a large oak may transpire up to 360 litres (dm^3) of water per hour on a sunny day! This is especially significant in areas of the world where there are large forests, creating a humid atmosphere. This water vapour eventually falls as rain to continue the cycle, but where large areas are de-forested this cycle is disturbed changing the humidity of the air and the subsequent rainfall.

The Transpiration Stream

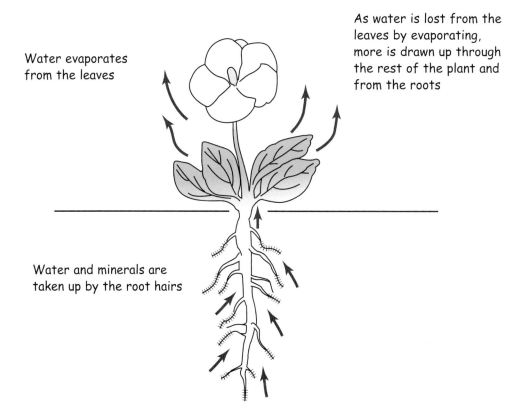

Water evaporates from the leaves

As water is lost from the leaves by evaporating, more is drawn up through the rest of the plant and from the roots

Water and minerals are taken up by the root hairs

Domestic water cycle

There is also a domestic water cycle, the average personal use is about 273 pints or 155 litres (dm^3) per day and most of it leaves our houses as waste water from bathrooms and kitchens and goes into the sewers. This, together with water from industry must be treated before it returns to the distribution system in a form suitable for drinking.

Graphing package

Domestic Water Usage

Taking a bath	40 litres
Shower	20 litres
Flushing the toilet	10 litres
Washing hands or face	9 litres
Drinking water	1 litre
Brushing teeth	1 litre
Dishwasher load	30 litres
Washing machine load	100 litres
Other… cooking etc.	30 litres

At the waste water treatment plant (sewage works), screening removes large debris and a series of filters removes other matter. The water then either goes to a water treatment plant or is allowed into the rivers and seas where it continues the cycle of evaporation, condensation and precipitation. Rain water collects in reservoirs, and together with water from under the ground and from rivers it goes to a water treatment plant where it is disinfected, stored and pumped through the main distribution system to our homes, hospitals and industry.

SKILLS

- Fair testing, dealing with variables, repeating procedures.
- Working cooperatively, using apparatus and observing accurately.
- Using a thermometer and a stopwatch.
- Measuring volume and perimeter.
- Constructing a line graph.

THE WATER CYCLE
Key ideas and activities

Different liquids will evaporate (at different rates) if left uncovered

After a discussion about puddles drying up, clothes drying and the introduction of the word evaporation, simple activities can be set up to demonstrate the idea, depending on the age and experience of the children. After these activities discuss where the liquid went. It is also worth discussing with the children the fact that the liquid does not need to be hot to turn to a gas/vapour.

(a) Leave out dishes of different liquids, eg water, vinegar, lemon juice, salty water and tea for the children to observe. Duplicate some and cover them. What happens to the uncovered dishes? Is anything left behind? Can you smell anything, why?

(b) The first part of the above activity may be carried out in a more quantitative way as an extension for the more able children or as an investigation, 'which liquid evaporates the fastest?' This will involve the idea of fair testing because the dishes, quantity of liquid, conditions in which they are left must remain the same. It is useful to calibrate the containers at the start or use calibrated beakers. The children can plan this investigation as a way of reminding them of the importance of fair testing and the variables involved. Older children can construct a line graph of the rate of evaporation of the different liquids, which would involve timing and measuring the liquids daily or more frequently depending on the temperature of the classroom and volume of liquid.

I predict water will evaporate the fastest.

Liquid used for test	Amount of liquid left and other observations				
	Monday	Tuesday	Wednesday	Thursday	Friday
lemon juice covered					
lemon juice uncovered					
vinegar covered					
vinegar uncovered					
salty water covered					
salty water uncovered					
water covered					
water uncovered					

THE WATER CYCLE RS•C

A graph to show the rate of evaporation of liquids.

Graphing package

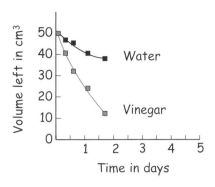

We could draw more than one line using different colours for different liquids.

(c) If it is possible and it has been raining (or fake it!) monitor a puddle in the school playground by drawing around the perimeter with chalk at regular intervals throughout the day as the water evaporates. Before it completely disappears get the children to predict where the next chalk line will come. Reconstruct on paper in the classroom.

Evaporation is affected by various factors

The following activities may be set up as directed classroom activities, but all allow the opportunity for excellent investigations to take place. Many children would say for example, that puddles dry up faster when it is sunny or warm, why? How would they prove that temperature affects evaporation, and is it so for liquids other than water? These activities also allow for the introduction or revision of 'variables' and fair testing and line graphs.

Datalogging

(a) Temperature Set up containers of liquids or just water in areas of varying temperature including the refrigerator and near radiators or the sunshine. Use wall thermometers to show the different temperatures of the different areas or leave thermometers or sensors in the liquids to monitor their temperature. The children can then record the different amounts of liquid.

Safety!
• Spirit thermometers should be used.
• Care when handling glassware.
• Care when handling hot and cold things.

(b) Surface area Investigate which evaporates faster, the same volume of water in a wide, shallow container or a narrow deep container. Set up a variety of containers for this activity. If they are jars the children can draw around the openings onto squared paper to calculate the surface areas. This is an activity that most would have done in maths. Square or rectangular plastic boxes may also be used, and the areas of these can be calculated by multiplying their length by the breadth (for the more able children).

Keep the test fair... same liquid, same temperature, same volume of liquid. Which do you think will be the fastest?

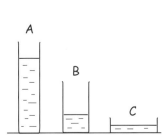

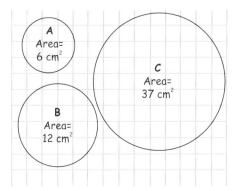

(c) **Wind factor** Some special equipment may be needed, an electric fan to supply the wind! This could be carried out as a demonstration using a small quantity of liquid in a shallow container standing in front of the fan and another behind it, just to demonstrate the point.

(d) **Everyday context** Set the above activity in an everyday context of washing drying on a line. Small 'washing lines' can be set up in different areas of the classroom, eg in front of the fan, over a radiator. Different groups of children can contribute identical sized but different types of material for each line. So, for example the 'cotton' group has identical sized pieces of cotton on different lines in the classroom. The pieces of material are wetted with the same volume of water. As there are two variables here, the type of material and classroom condition, this maybe more suitable for the more able children. Other children could carry out this activity with one variable only.

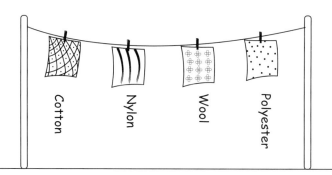

Make the pieces of fabric the same size and put the lines in different places.

Discuss the results and get the children to identify various situations in an everyday context where evaporation occurs, include the use of drying equipment such as hairdryers and tumble dryers.

Evaporation can be used as a way of recovering dissolved solids

This can be very simply introduced as shown in activity (a) as a basic principle by including salty water as one of the liquids left to evaporate. This can then be discussed along with the question, what would happen if you left out a sugar solution? The children could then prepare a series of solutions eg coffee, sugar, bicarbonate of soda and evaporate the water off. (This is further explored in the section on Separating mixtures of materials).

The concept of water vapour condensing to water as it cools needs to be introduced either at this stage or before. (Refer also to section on Heating and cooling materials, kettle experiment.)

Condensation and evaporation as part of the water cycle

(a) **Water vapour can condense to form water and this is the reverse of evaporation** This can either be a whole class or group demonstration. Set up a bowl of warm water, cover with a dish containing ice. As the water evaporates and hits the cold saucer, it condenses and 'rains'.

(b) Repeat the activity using cooler water, which just takes longer to work. Discuss this with the children. Water does not need to be

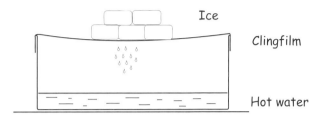

What happens if we use cold water and no ice?

hot to evaporate and condense, but the process is faster with hot water.

(c) Small groups of children can have their own screw top jar with ice inside. After about 5 minutes condensation is seen on the outside of the jar, where water vapour in the air has cooled against the cold glass.

Some children have great difficulty with the concept of condensation and are convinced that the ice melts and goes through the jar or the saucer. Try using a very cold glass straight from the refrigerator and eliminate the ice.

You could use the concept cartoon *Ice cream* here.

(d) **Plants are part of the water cycle, water evaporates from their leaves** A demonstration model can be set up to show that water evaporates from the leaves by covering a stem/leaf section of a pot plant with a small polythene bag. Water evaporates from the leaves and condenses on the inside of the bag. Also a mini-cycle can be set up, by enclosing the whole plant in a clear polythene bag. Water evaporates from the soil and the leaf surface, condenses on the inside of the bag and drips back onto the plant.

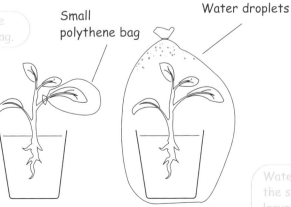

Enclose part of the plant in a plastic bag.

Water evaporates from the surface of the leaves (and soil). It condenses on the inside of the bag and drips back onto the plants.

(e) **Water evaporates from the Earth's surface, condenses to form clouds and falls as rain** Discuss all the previous activities in the wider context of the water cycle and the weather. A demonstration model can be set up to simulate this using an aquarium with land and water, cover with cling film and put ice in a saucer on the top, which can be renewed. Use this as a model to get the children to draw their own water cycle from what is happening. Then draw the water cycle on a larger scale in real terms.

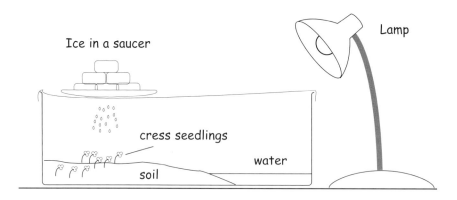

The Water Cycle

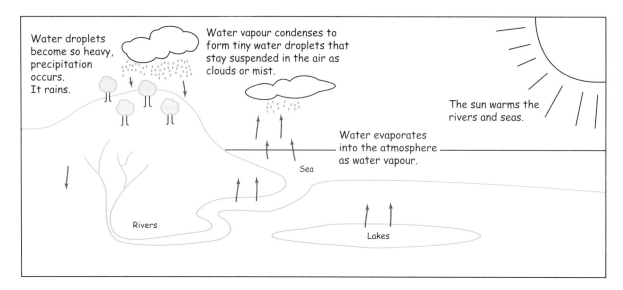

(a) **How much water do we use or waste?** It is worth making the children aware of how much water we use, especially since the amount of available water is a life-threatening issue in many countries. Referring to the 'Usage table' page 105, get the older children to calculate how much water they or their family use in a week. They can record this as a chart.

The domestic water cycle

(b) **Where does water come from and our waste go?** Older children could research this for themselves using books or a CD-ROM. Alternatively, they can make a pictorial cycle from a list of sentences you give them about the domestic cycle. Local water boards will often give free information leaflets or visit the school to give a talk.

by Michael Rosen

Salty Wave Salty, sea, water evaporates leaving the salt behind on your skin.

Chippy Breath Condensation occurs when water vapour hits a cold surface and changes to a liquid. See page 84.

Steamy Shower Hot water evaporates, then condenses in the bathroom, dripping of the ceiling. See page 85.

Hot Pants The hot air in a tumble drier speeds up evaporation making the clothes dry quickly. See page 85.

Grandad's Airer Water evaporates from the clothes on an airer, making them dry. See page 85.

Thirsty Land The sun's heat makes the water evaporate from the land, drying up the land. See page 85.

Salty Wave

When I went through the wave I gobbled a great gulpful of salt-water.

That night in bed my hand nestling on my neck found tiny crusts; white grains.

And when I licked my finger-tip, it was as salty as a crisp – or a wave.

ICE CREAM

CONCEPT CARTOONS

Ice cream

Although the children will have experience of condensation they are unlikely to have well-formed ideas about where the condensed water comes from. The concept cartoon invites them to consider and investigate a number of possibilities, and they may well think of other possibilities themselves. The fact that the condensation comes from the air may appear to be the least likely possibility to many of the children. Wrapping the ice cream tub in polythene or aluminium foil and observing where the condensation forms should help to clarify their ideas. Investigations such as this help to lay the foundation for later work on the structure of matter and conservation of mass.

this page has been intentionally left blank for your notes

SEPARATING MIXTURES

Science background for teachers

VOCABULARY

Mixture, solution, solid, liquid, separating, sieving, filtering, decanting, evaporating, condensing, distilling, chemical, industry, retrieved, process, vapour, sediment, soluble, insoluble

A mixture is formed when two or more substances are mixed physically and not chemically combined. These substances can be separated and recovered again by physical and not chemical means, although not always easily. The method of separating may depend on the sizes of the particles involved or on their physical properties. Examples may include sieving, decanting, filtering, evaporating, distilling and chromatography.

Very young children experience sieving when they play with sand and water and all children see it in everyday contexts with tea strainers, tea bags, coffee pots and straining water and vegetables with a sieve or colander.

Sieving involves smaller particles in a mixture passing through holes, whilst the larger particles remain behind in the sieve. Sieves of different grades can be used to separate out particles of different sizes. Anything with holes or slits can be used to sieve. Children can invent their own sieves to suit their purpose.

Straining involves separating solid in liquid mixtures where the solid particles are large, such as vegetables in water, where you want to retrieve the solid.

Filtering is generally used to separate liquid from smaller solid particles where the holes in the smallest of strainers are still too large. Filtering uses special filter paper, which allows the liquid through but catches the very small solid particles. Other materials such as gravel and charcoal may also be used as filters to trap tiny solid particles. They are used in industries such as the water industry for cleaning wastewater where paper is obviously not practical.

Leaving a mixture to stand then pouring the liquid carefully off is called decanting. This may be done to wines to leave behind the sediment, but is not often mentioned and is a valuable process to demonstrate with children. This may be seen in the traditional coffee-making method (not instant) where the coffee and hot water are left in a jug to stand and the coffee is decanted off, similarly with making tea in a teapot. The coffee grounds and tea leaves sink to the bottom in a matter of minutes and a strainer is only needed as the pot is tipped more. If a mixture of soil and water are left to stand, the water that is obtained by pouring off the top layer is very clear, but some water remains behind in the soil, which can be filtered. Children can experiment with a variety of different filters, some of which are used in industry eg gravel and sand.

Tip wash the gravel/sand first or they make the water dirtier!

Magnetic materials (nickel, cobalt and iron including its alloys such as steel) may also be separated from non-magnetic materials such as plastics or aluminium using a magnet. This is useful in the scrap metal industry. Take care when choosing metal objects to use, many these days contain steel and are coated with a brass lacquer, such as brass tacks. They are magnetic even though brass is not.

Evaporating the liquid

The filtering process will not separate a true solution because the solid breaks down into particles so small they will pass through filter paper and these are dispersed evenly in the liquid. The particles of the solid then form weak interactions with the molecules of the liquid. If the solution is heated, the heat energy is enough to break the interactions. The liquid evaporates leaving the solid behind. This will also happen without heat, but takes much longer. This will retrieve the solid, but the liquid will be 'lost' as a vapour. To recover the liquid, the vapour is collected, cooled and condensed and this process is called distillation. In countries where there is a shortage of clean drinking water, this may be obtained by heating seawater, the vapour is then cooled and the result is pure water. The recovered salt is also used. Salt is obtained on a large scale from either evaporating seawater or from the ground. Sea salt is used primarily for cooking and is produced in relatively small quantities. The chemical industry uses salt in larger quantities and obtains it from the ground in one of two ways. It is either mined by cutting, drilling and blasting the rock or more commonly by pumping water down into the rock where it dissolves and is brought to the surface as brine (salty water).

Chromatography This literally means 'writing with colour' and is a technique used to separate solutions where more than one solute (solid) is dissolved in the solvent (liquid). For example colours and dyes are often made up of various components to obtain one colour. These can be separated out into the different components using chromatography. A spot of the solution to be separated is put onto absorbent paper, a solvent (water if the solute is water-soluble) is allowed to be absorbed onto the paper. As it spreads through the paper the components of the solution separate out through the paper at different speeds because they are attracted to the paper fibres to different degrees. Thus the colour components of a dye will separate out. Sometimes the components will not separate by this method. This may be because they are not soluble in the solvent that is being used, it is necessary to test them before the children try the activity. Not all water-soluble inks and fibre tip pens will separate into component colours using only water as the solvent. Many of the colours in a tube of 'Smarties', often used in the past for this activity, will no longer separate out with water! Water-soluble drawing inks and overhead projector pens work well.

SKILLS
- Working cooperatively.
- Using equipment with care.
- Observing with care and accuracy.
- Recording results accurately.
- Understanding a fair test.

SEPARATING MIXTURES
Key ideas and activities

These activities begin with those for young children and progress through to the end of the primary stage;

Particles of different sizes can be separated by sieving

(a) A variety of different solid/solid mixtures can be given to children together with sieves and colanders of different sizes so that they can separate out the different sized pieces. It is better to try and make these fit into everyday situations rather than make them too contrived. Who needs to separate sand from peas? A few suggestions are: sieving lumps from flour, rice from salt, coffee from coffee beans, stones from soil, stones from sand, different sized seeds, sugar from sugar lumps, buttons in a button box, buttons and poppers, beads. In the last activities, the children could make their own sieve if you can't find one that is big enough. You can cut larger holes on the mesh on a cheap, large, plastic sieve or cut holes out of the bottom of a card box. The children may have ideas of their own.

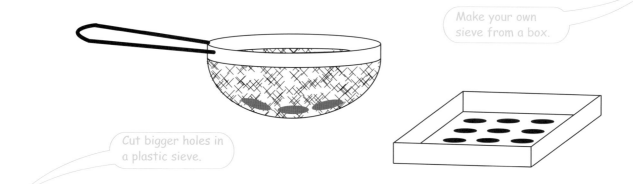

Make your own sieve from a box.

Cut bigger holes in a plastic sieve.

(b) Use graded sieves to separate soil samples, so that the composition of the soil can be closely observed. (Refer also to section on Rocks and Soils.)

Insoluble solids can be separated from liquids by filtering (and straining)

(a) Discuss how to and try to separate a variety of liquid/solid mixtures that are within the children's everyday experience, such as water with different vegetables, washing rice, straining tea, looking at a cafetière. Are you retrieving the solid or the liquid?

(b) Look at teabags, put them in hot water and get the children to observe and explain what is happening. Why is the resultant liquid brown? This activity which is familiar to the children, can lead into the idea of 'filters'.

(c) Demonstrate making coffee with a coffee filter, what happens if you use a fine strainer instead? Why are different grades of coffee recommended for different types of coffee machines?

(d) Recover the sand and water from a sand and water mixture. Try a fine sieve first, the water is still very cloudy even though much of the sand has been retrieved. If the children then continue to filter the cloudy water using filter paper in a funnel, the water will get clearer.

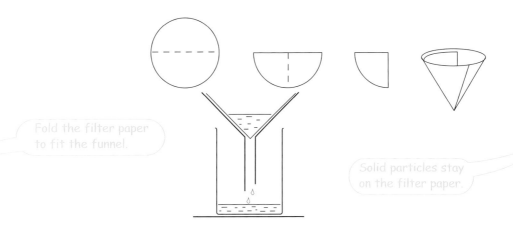

Fold the filter paper to fit the funnel.

Solid particles stay on the filter paper.

(e) Discuss filters with the children and talk about uses for them other than domestic uses, such as cleaning wastewater. As it is not always practical to use filter paper (it easily tears when wet as they may discover), discuss the use of other filters. A good system for trying out other filters other than with a filter funnel is to use a plastic lemonade bottle. Cut off the top at about a third of the way down. Put muslin or nylon secured with an elastic band over the narrow, neck end and invert this into the base of the bottle. Now you can put any filter material into the top. It is easy and cheap to make.

Safety!

- Care when handling glassware.
- Care with candles; stand in a sand-filled tray.
- Care with hot liquids.
- Care when using soils and sand.
- Plastic gloves should be used and hands washed afterwards.
- The soil should be checked for sharp materials and animal faeces.

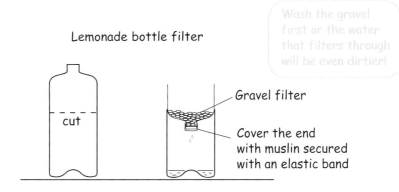

Lemonade bottle filter

Wash the gravel first or the water that filters through will be even dirtier!

Gravel filter
Cover the end with muslin secured with an elastic band

(f) **Investigate** 'Which is the best filter?' You may get the children to suggest different filters to try or offer them a range, such as washed gravel, washed sand, cotton wool, steel wool, muslin or nylon tights. Get them to predict, with reasons, which they think will be the best. Some children might repeat the experiment trying two filters, either one after the other or putting two filters or more together in layers, one on top of another, to get the best result.

(g) **Decanting** Leave a jar of muddy water to stand for the children to see for the day, then pour off the water as far as you can without disturbing the sediment. The separated water is usually very clear.

SEPARATING MIXTURES

Recovering dissolved solids by evaporation

(a) The children need to have experience of evaporating and may have experience of solids being left behind when the liquid evaporates from the evaporating activity (Refer also to Water cycle-evaporation section.) This can be discussed with them. Let the children make their own solutions and leave them out in shallow dishes for them to observe the solid that is left behind eg salt water, sugar solution, ink, instant coffee.

(b) The above activity (a) can be repeated but speeded up by gently heating the solution over a candle. Get them to predict the colour of the ink and coffee vapour before they start.

(c) Discuss where salt comes from and get them to plan an investigation into recovering clean salt from either dirty, salty water or for older, more able children dirty, gritty salt. Both investigations require filtration and evaporation, but for the second activity the children need to realise that water must be added first to make a solution

Using knowledge of solids, liquids and gases to decide how mixtures might be separated

Recovering the liquid from a solution: Investigate how to obtain pure water from salt water, a problem that often has to be solved in countries where there is a water shortage. This is for older or more able children to investigate, or it can be a class demonstration. The basic principle involved is evaporating the water from the salt and water mixture and then condensing the vapour to obtain the clean liquid, leaving the salt behind. Children need to be familiar with the concept of condensation, discuss and remind them. The problems of obtaining fresh water in countries where there is little available can also be discussed. (The investigation may be made more complex by having sandy, salty water, which needs to be filtered first to remove the sand.) This is how it may be demonstrated.

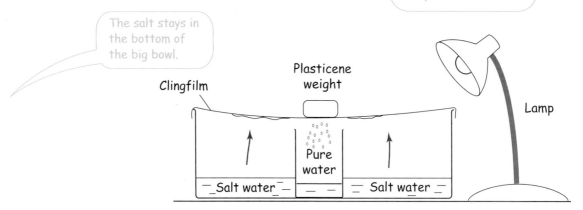

Recovering pure water from salt water

You could use the concept cartoon *Sugar in tea* here.

Chromatography Test a variety of water soluble inks or fibre tip pens before giving them to the children to make sure that the colours will separate out. There are two basic methods:

- Cut a strip of filter paper, put a dot of the colour or colours to be tested about 3 cm up from the bottom and stand in a tall jar containing water that is approximately 2 cm in depth. The water seeps up the paper, washes through the colour and separates out the colours.

- Put a dot of colour in the centre of a circular piece of filter paper. Cut 2 slits in the circle from the edge to the spot and bend the slit up, making a 'wick'. Allow the 'wick' to dip into water so that the water seeps up to the ink spot. The colours will then separate out towards the circumference of the paper.

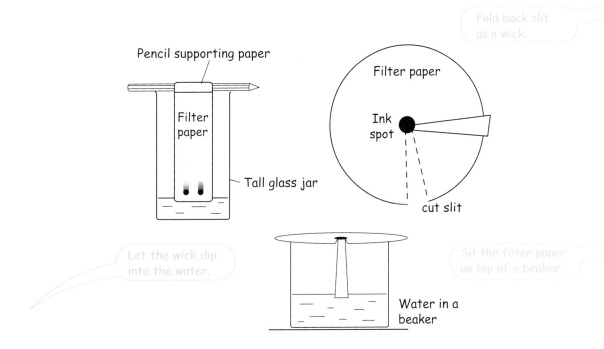

Separating magnetic materials

The children may need to be reminded about earlier activities with magnetic and non-magnetic materials.

(a) **Separating metal (magnetic) from other material**
Separate the metal poppers or buttons from the plastic ones in the button box. A very simple investigation for younger children.

(b) **Separating a mixture of metals** Using a magnet, separate the brass from the steel screws that have been mixed in a box. Discuss the separation of metals for recycling for example at waste disposal facilities.

by Michael Rosen

Soapy Spuds Potatoes can end up in the sink when you strain them!

Button Box Anything with holes or gaps can be a sieve, even fingers.

Summer Sand When you wash holiday clothes, sometimes the sand stays in the machine. See page 71.

Salty Wave When you swim in the sea, sometimes the salt stays on your skin. See page 111.

Soapy Spuds

Dad was straining the potatoes
when he said to me:
Ssssh!
Don't tell anyone:
some of the spuds have fallen out of the colander
and into the sink.
I've put the ones that fell out
back in with the others.
No one'll notice.
Sssh!
Don't tell anyone.

So I didn't tell anyone.
And he was right.
No one noticed.
No one at all.
He got away with it.
Brilliant.

One problem:
my potatoes
tasted of washing up.

Thanks Dad.

Button Box

I go fishing
in the button box
digging my hand
deep down inside,
making a cup
with my fingers,
pulling my hand
slowly upwards,
letting the small boring buttons
slip through the gaps
between my fingers…

…so I can catch
the gold buckle
off grandma's shoe,
the fat domed button
that comes dressed
in purple cloth,
the American Army badge,
and the glass eye-ball
that stares at me
from the palm of my hand.

SUGAR IN TEA

SEPARATING MIXTURES

CONCEPT CARTOONS

Sugar in tea

This concept cartoon invites the children to reverse the familiar process of dissolving. It also challenges the children's ideas about what happens to the sugar in the tea – does it disappear completely as it dissolves or can it be recovered from the tea? The children can investigate the possibilities shown in the concept carton as well as other possibilities that they might suggest. Salt is a useful alternative, since it can be separated more easily from water than the sugar. Other means of separating materials would be useful ways to follow up this investigation.